Getting Started with
HAM RADIO

A Guide to your FIRST
Amateur Radio Station

By Steve Ford, WB8IMY

Published by:

ARRL *The national association for*
AMATEUR RADIO

225 Main Street • Newington, CT 06111-1494

www.arrl.org

Foreword

If you're a new ham, you've been patient for a long time. You've struggled through studies and examinations, only to wait while the FCC processed your documents and awarded that precious Amateur Radio call sign.

But now you finally have it and you're itching to put the license to work. Suddenly, a flurry of questions arises out of nowhere…

- Which transceiver should I buy? What about an antenna?

- Which mode should I use? SSB, CW or any one of the many digital modes?

- How do I hook up all this hardware?

- How do I make contacts with my chosen mode?

If you're lucky, you may have an *Elmer* to help you. In Amateur Radio tradition, an Elmer is a knowledgeable ham who can answer all your questions and help you avoid serious pitfalls as you begin your journey. But if you don't have the luxury of an Elmer (many amateurs these days do not), what can you do?

Think of this book as your "paper Elmer." It doesn't contain every fact in the Amateur Radio universe, but it contains the most important information you'll need to get started on the right foot. As you breeze through the book, you'll take a guided tour of much that ham radio has to offer. Pick a facet that interests you and you'll have all the practical advice you need at your fingertips. This is the kind of how-to book that never goes out of style.

Welcome to Amateur Radio!

David Sumner, K1ZZ
Executive Vice President
Newington, Connecticut
July 2006

Table of Contents

The Benefits of Membership

Welcome to ARRL

This is your personal invitation to join the only national organization representing Amateur Radio today – the American Radio Relay League. For nearly a century, the ARRL has been the official voice of and for the radio amateur. The ARRL is here to provide you with all the tools you need to ensure your Amateur Radio experience is all it can be.

Membership in ARRL provides you with countless benefits. As a member, you will receive our monthly journal QST, e-letters and have access to technical assistance from our experts on all aspects of Amateur Radio. You can hear up to date information from headquarters station W1AW with daily transmissions of bulletins as well as Morse code practice sessions. The ARRL also coordinates an extensive field organization that includes volunteers who provide technical information for radio amateurs, sponsor local special events and public-service activities, and serve as the League's eyes and ears on the local level. In addition, ARRL represents US amateurs with the Federal Communications Commission and other government agencies in the US and abroad, working to ensure the frequencies that you enjoy will always be available. And these are just some of the reasons to join ARRL. When you look at all the benefits, you'll realize that you can't get a better value anywhere!

There are so many reasons to join us here at the ARRL. Regardless of your Amateur Radio interest, ARRL membership is relevant and important. We hope you enjoy this publication and your experience with Amateur Radio. Visit **www.arrl.org/join** *and join today! And please don't hesitate to contact us if you have any questions.*

ARRL *The national association for* **AMATEUR RADIO**
225 Main Street
Newington, CT 06111-1494 USA

1 Your First Radio

Your Amateur Radio station reflects your interests and, naturally, it will change as you explore different facets of the hobby.

But you have to start somewhere, right?

Fortunately, there are some fundamentals that all Amateur Radio stations have in common. Once you understand the basics, you can apply them to your first station and every station thereafter.

See **Figure 1.1**. The three building blocks of every ham station are:

- Power Supply
- Transceiver
- Antenna System

In this chapter we'll discuss the first two building blocks. Antennas deserve a chapter all their own!

THE POWER SUPPLY

Without a power supply, a transceiver is a lifeless hunk of metal and plastic. The power supply supplies the "juice" that makes ham radio possible.

If you're considering a handheld transceiver for use on VHF or UHF FM, you'll be pleased to know that most of these radios come with their own rechargeable batteries. But if you want to operate the radio

This MFJ power supply is a switching model that supplies substantial current in a small, lightweight package.

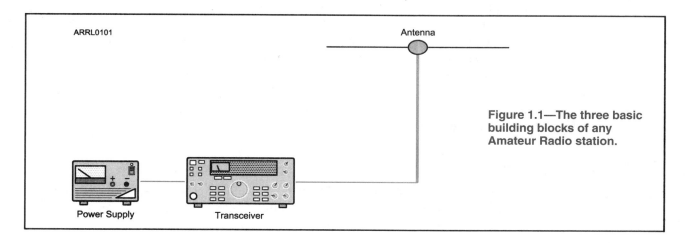

Figure 1.1—The three basic building blocks of any Amateur Radio station.

without the battery (in your house or apartment, for example), you may want to invest in a small dc power supply—13.8 volts (V) with a current capacity of about 3 amps (A) will do the job nicely. You can find these at retailers such as RadioShack for about $40 or less. With a dc power supply, you won't have to worry about your battery running down when you are in the middle of a conversation.

As you step up to larger radios with more output power, you'll need larger power supplies to run them. Most of these transceivers do not have their own power supplies, so read their specifications before you buy. A transceiver with a maximum output power of 100 W will require about 25 A of current at 13.8 V when you are operating the radio at "full throttle." That kind of power supply will set you back about $100 to $200, depending on the overall design.

> **Don't worry about buying a power supply with too much current capacity. Your equipment will only draw the current it needs—no more, no less.**

Don't worry about buying a power supply with too much current capacity. Your equipment will only draw the current it needs—no more, no less. In fact, it is probably safe to say that you can never have too much current capacity. It may seem economically foolish to invest $200 in a 25-A power supply when all you want to power is a 5-W handheld radio. However, if you think you'll be upgrading to a larger radio in the near future, you may want to get the big power supply today (especially if you find a great deal on a high-current supply).

When shopping for a power supply, beware of one potential stumbling block. Power supplies are often rated by their *continuous* and *intermittent* (ICS) current capacities. The figure you want to look at is the *continuous* rating—the amount of current the power supply can provide continuously. Don't be misled by an advertisement that promises a fantastic deal on, say, a 30-A supply. Are those 30 amps provided intermittently—only for short periods of time—or continuously? You need continuous power, so check and be sure!

It is also worth mentioning that you'll find two types of ham-grade power supplies for sale. The *linear* design uses a hefty transformer to shift the 120 V ac line voltage from your wall outlet to a lower voltage for later conversion to 13.8 V dc. These power supplies tend to be large and heavy, especially the high-current models.

Another approach to the power supply problem is the *switching* design. In the switching power supply, the ac line voltage is converted directly to dc and filtered. This high-voltage dc is then fed to a power oscillator that "switches" it on an off at a rate of about 20 to 500 kHz. The result is pulsating dc that can be applied to a transformer for conversion to 13.8 V or whatever is needed. The reason for doing this is that rapidly pulsating dc can be transformed to lower voltages without the need for large transformers. It is the transformer that accounts for most of the weight, size and cost of traditional linear power supplies. A switching power supply is much smaller and lighter, and usually less expensive.

Switching power supplies are the same type found in your computer and they are becoming more popular in Amateur Radio. The disadvantage of the switching supply is that some designs generate interfering signals that you can hear in your radio. If you're considering a switching power supply, look for models that boast low "RFI" (radio frequency interference). *QST* magazine occasionally reviews and tests switching power supplies. If you are an ARRL member, you can read previously published *QST* reviews on the ARRLWeb (**www.arrl.org**).

TRANSCEIVERS

We could spend every page of this book talking about transceivers. There are so many possibilities to consider, it boggles the mind. Still, there are some common guidelines that apply. Let's break them down the choices according to transceiver type.

VHF/UHF Handheld

Handheld transceivers (often called "HTs," which is a Motorola trade name) are almost exclusively for FM operating, usually with *repeaters* (we'll discuss repeaters in a later chapter). The strength of the handheld is its portability. You can clip a handheld to your belt, or slide it into your pocket, and go anywhere. Modern handhelds often include wide-range receivers, allowing you to monitor frequencies outside the ham bands such as police, fire, aircraft and more. Many handhelds also offer multiband operation (2 meters and 70 cm, for instance). And finally, handhelds are attractively priced. Among new transceivers, you'll spend the *least* amount of money on a handheld.

On the down side, handhelds don't offer much transmitter output power and their flexible antennas (known as "rubber duckies") are notoriously poor. Unless you have a sensitive repeater in the area to relay your signal, your range will probably be quite limited.

Yes, you can attach better antennas to handhelds, and you can even add an amplifier to increase your output power. However, what you end up with is a portable radio that isn't so portable anymore.

A typical handheld FM transceiver.

VHF/UHF Base or Mobile

The next step up the FM ladder is the mobile transceiver. These radios are compact, but not easily portable like a handheld. They usually provide much higher output power, often on the order of 50 W or more. Higher power is critical for good distance coverage, particularly with FM.

Mobile transceivers can span a wide price range, depending on the features they offer. You can pick up a basic single-band FM mobile transceiver for about $150 at the time of this writing. A dual-band FM mobile sells at about double the price.

Some mobile radios have special features such as digital capability, the ability to separate the body of the radio from the front panel (a nice feature for tight mobile environments) and more. Of course, these radios will be priced accordingly. If you don't require portability and don't mind paying a bit more, a mobile FM transceiver is a good choice.

The definition of a "base" transceiver has become blurred over the years. Strictly speaking, it means a transceiver intended for use inside a building. With today's compact technology, however, base transceivers can go mobile and mobile transceivers can be base radios. When a radio is in a vehicle, the vehicle's electrical

The Alinco DR-635T is a compact dual-band VHF FM mobile transceiver.

Mobile Installation Tips

● The equipment should be placed so that operation will not interfere with driving. Driving safely is always the primary consideration; operating radio equipment is secondary.

● Mount the radio securely. Never lay it on the seat beside you. If you get into an accident, that loose radio may become a deadly missile.

● If a piece of equipment will draw more than a few amps, it is best to run a heavy power cable directly to the battery. Don't try to power a 100-W radio from a cigarette lighter socket.

● Adequate and well-placed fuses are necessary to prevent fire hazards. For maximum safety, fuse both the hot (positive) and ground lines near where they attach to the battery.

system supplies the power. You can take that same radio indoors and use it as well, as long as you have the power supply we discussed earlier. In other words, mobile and base radios are almost entirely interchangeable.

For housebound VHF operating, the type of transceiver you choose has much to do with how you want to operate—now and in the future. Any mobile radio will do a fine job as a base rig for FM operating through repeaters or directly (known as *simplex*). If you want to try your hand at so-called "weak signal" work, you will need a transceiver that can do SSB and CW in addition to FM. These are often referred to as *multimode* VHF/UHF transceivers. (We'll discuss weak-signal operating in a later chapter.) A VHF/UHF multimode transceiver will cost in the neighborhood of $1700, depending on the model. There is a multimode VHF/UHF alternative, however, which we'll discuss in a moment. Read on!

HF Only

Remember that the High Frequency or *HF* bands are defined as those groups of frequencies from 1.8 to 30 MHz. (Technically speaking, 1.8 MHz is in the Medium Frequency or MF region, but we won't argue definitions here!) These are the most popular Amateur Radio bands because they can be used to communicate throughout the world at any time of the day or night.

Hams use several different modes of communication on these bands:

■ SSB – Single Sideband voice (the most popular voice mode)
■ CW – Continuous Wave on/off keying using Morse code telegraphy
■ Digital – Data communication using a variety of methods
■ AM – Amplitude Modulated voice
■ FM – Frequency Modulated voice (this only takes place on the high end of the HF spectrum from 29.5 to 30 MHz)

Is QRP Right for You?

Low power *QRP* operating is a thriving part of the Amateur Radio scene. QRP enthusiasts operate at only 5 watts output or less using primarily CW, but they also use digital modes and occasionally voice.

The great advantage of QRP is cost. A QRP transceiver built from a kit can cost less than $200. Low power consumption is another major plus. QRP transceivers can be easily powered from batteries, which make them great for outdoor or emergency operating.

The disadvantage of QRP is that you need a very good antenna to make contacts with reasonable ease. With such low output power you must compensate at the antenna to make yourself heard. This isn't to say that you can't make QRP contacts with a poorer antenna (such as a small mobile antenna), but it will be much more difficult.

You'll find that most HF-only transceivers support all of these modes, although some radios lack FM and a few do not include AM. Most HF radios offer transmit coverage of all HF ham bands, plus the ability to listen to all frequencies throughout the HF range. This general-coverage receive capability is handy because it allows you to eavesdrop on shortwave broadcasts and other interesting signals.

HF-only transceivers typically offer 100 W output. With a decent antenna, this is enough power to make global contacts when the bands are open. You'll also notice a few HF transceivers designed

Any mobile radio will do a fine job as a base rig for FM operating through repeaters or directly (known as simplex).

for low-power operating, better known as *QRP*. These transceivers are available in multimode, multiband models as well as single band, single mode units (usually CW only). QRP transceivers are often less expensive and physically smaller that their high-power cousins, but QRP hamming is an art unto itself and requires certain skills to be successful. See the sidebar "Is QRP for You?"

This Kenwood TS-2000 transceiver is a so-called "dc to daylight" radio that operates from 1.8 to 440 MHz. It also includes special features for satellite operating.

"DC to Daylight"

The transceiver trend in recent years has been away from HF-only radios. In the 1990s ICOM introduced the IC-706, the first multimode transceiver that spanned the HF bands *and* the 6- and 2-meter VHF bands. Other manufacturers followed suit and now you'll find transceivers that cover 1.8 to 54 MHz and even 1.8 to 450 MHz in a single box. These so-called "dc to daylight" transceivers are among the most popular radios sold today. It's easy to see why. They offer the ability to enjoy the global coverage of the HF bands while opening the door to the enjoyment of VHF and UHF. (This is the "multimode VHF/UHF alternative" mentioned earlier.)

Many of these dc-to-daylight transceivers are also quite compact, serving as either mobile or base radios. Others offer extended features such as the ability to function as amateur satellite transceivers.

Which Transceiver is Right for You?

The answer to this question depends on how much money you have to spend and what you hope to do with the radio once you buy it. You can spend as little as $200 and as much as $13,000!

When in doubt, simplify. It's time for a table…

Transceiver	Pro	Con
Handheld VHF/UHF FM	Inexpensive; small	Limited range
VHF/UHF FM Mobile	Great for local FM work, base or mobile	Bulkier than a handheld; more expensive, too
HF only; "low end" ($500 - $700)	Powerful, yet moderately priced	Limited features; mediocre receiver performance
HF + 6 meters; "high end" ($2000+)	Excellent receiver and many features	Expensive and bulky
DC to Daylight ($700 to $1700)	Every band and mode you'll probably ever want	Receiver quality can vary; pricey for a starter radio
QRP (Less than $600)	Inexpensive; small	Often single band/mode; low output power can be a challenge. Some radios are CW only.

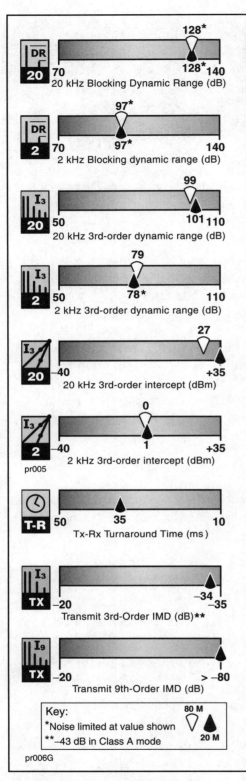

Figure 1.2—Beginning in 2006, *QST* magazine began publishing these key performance tables in every transceiver review. You can use these graphs to quickly understand how well a particular radio performed under ARRL Laboratory testing.

When reading this table, you'll likely notice the comments concerning receiver performance. When it comes to transceivers, receiver performance is critical to your operating enjoyment. For VHF/UHF FM transceivers, you need a receiver that is sensitive enough to hear local and distant stations, yet selective enough to block interference. In VHF/UHF applications, this interference often comes from nearby commercial transmitters such as police and fire communications and paging systems. You'll want to make sure that your FM radio has enough selectivity to stop these signals from getting all the way to your speaker or headphones.

On HF the basic receiver requirements are much the same, but interference is usually worse and the desired signals can be much weaker. Combine this with noise from Mother Nature (static crashes) and you have a challenging reception environment indeed! For casual chit-chat operating, just about any HF transceiver will do the job. It's when you begin hunting for weak signals, or operating in contests, that receiver performance becomes particularly important.

The ancient axiom "you get what you pay for" is as true in Amateur Radio as it is anywhere else. You can save a bundle of money with a low-end transceiver, but don't expect high-end receiver performance.

The best approach to sorting out your transceiver options is to become a member of the ARRL so that you'll receive *QST* magazine each month with its detailed Product Reviews (and have access to the ARRL online Product Review archives). By carefully studying the *QST* reviews you can make an informed choice, getting maximum "bang" for your buck. Even if you don't fully understand the Product Review measurement results, look for the key results table (see **Figure 1.2**). This graphic will tell you how the radio performed in all the areas that matter most.

BELLS AND WHISTLES

Every transceiver has a collection of features—some more important than others. The number of features directly affects the price you'll pay. More features = more money. In ham slang, features are known as "bells and whistles."

Let's talk about several common features and rate their overall *operating enjoyment factor*…

■ **Noise Blanker**

Operating Enjoyment Factor: *High*

Noise blankers are your best friends when you're doing battle with local noise sources, especially rhythmic noises such as the sharp *putt-putt-putts* of engine spark plugs. A well designed noise blanker can make this annoying noise disappear. Good noise blankers are critical for HF mobile operating.

■ Audio DSP

Operating Enjoyment Factor: *Medium*

Audio DSP means digital signal processing at the audio output stage of the transceiver. Audio DSP can significantly reduce noise and many DSP designs can also eliminate the squeal you'll hear on the HF bands when an interfering station transmits a continuous unmodulated signal on the frequency you are receiving.

■ IF DSP

Operating Enjoyment Factor: *High*

Digital signal processing at the receiver's intermediate frequency (IF) is something you'll find in moderately priced transceivers (or higher). Don't confuse this with audio-stage DSP.

Lower-cost transceivers use mechanical or crystal filters in the IF stage to reduce the receiver bandwidth and eliminate interference in a fixed range of frequencies. These filters tend to be expensive (as much as $200 each) and you have to purchase them at an additional cost. In contrast, DSP IF filters are provided as standard equipment; you don't need to buy them later. Better yet, they can be adjusted to achieve whatever bandwidth you desire.

■ Speech Processing

Operator Enjoyment Factor: *Low*

Speech processing, sometimes called "speech compression," is a method of boosting the average output power in an SSB voice signal. It can make a difference if you are using low power and you're trying to make yourself heard through interference or poor band conditions. However, speech processing can also introduce distortion if not used properly. This is a nice-to-have feature if you are into SSB, but not strictly necessary.

■ IF Shift

Operating Enjoyment Factor: *Medium*

The IF Shift function allows you to change the receiver's intermediate frequency slightly to "move" it off an interfering signal. If you're into CW, IF Shift can be

> *Speech processing, sometimes called "speech compression," is a method of boosting the average output power in an SSB voice signal.*

quite useful in crowded conditions, but its benefits on SSB are less pronounced.

■ Built-in SWR Meter

Operating Enjoyment Factor: *High*

As you'll learn in the next chapter, knowing your antenna system SWR is critical. SWR—Standing Wave Ratio—is an indicator of how well your antenna system is matched to your radio. A poor match, manifest as a high SWR, will cause your transceiver to reduce its output and may even result in damage to your equipment. Many hams purchase outboard SWR meters as separate station accessories, but if your radio already includes the meter, so much the better.

■ Built-in Automatic Antenna Tuner

Operating Enjoyment Factor: *Low*

Depending on the type of antenna system you use, this feature can be either extremely useful or totally useless. Built-in tuners work well when you need to match your radio to an antenna system with a reasonably low SWR (less than 3:1). For antenna systems with high SWRs, they offer little benefit.

■ Computer Interface

Operating Enjoyment Factor: *High*

Computers and transceivers go hand-in-hand these days. If you want to be able to tie your radio and computer together (for memory programming, automatic frequency logging, etc), look for this feature. *Beware:* Some transceivers include the complete interface while others only make it available as an option at an additional price.

■ Memory Channels

Operating Enjoyment Factor: *High*

A well-designed transceiver memory system allows you to store dozens (sometimes hundreds!) of frequencies for later recall, along with power settings and much more. *Tip:* Look for transceivers that provide alphanumeric "tags" for each memory slot. This feature is great because it lets you assign short strings of letters or numbers that will appear on the transceiver's screen as you scroll through the memories. (Example: SOUTH REPEATER.)

■ Voice Operated Switch (VOX)

Operating Enjoyment Factor: *Low*

If you are a voice operator, VOX can be useful because it frees you from having to repeatedly press the TRANSMIT/RECEIVE switch. Whenever you stop talking, VOX will switch your transceiver back to receive automatically. When you speak again, VOX will place your radio in the transmit mode.

The downside of VOX is that it can be annoying—both to you and the listener. It will do its duty every time you pause to consider a thought, popping the radio rapidly from transmit to receive and back again. This causes the "ahh" effect—the tendency for VOX users to say "ahhhhhh" to keep their radios transmitting while they think of what they want to say next!

■ CW Keyer

Operating Enjoyment Factor: *High*

If you are into CW operating, this is an important feature. Many CW operators use "paddles" that work with CW keyers to automatically generate dots and dashes whenever you touch them. You can buy external CW keyers, but having the keyer already built into the radio is highly desirable.

WHAT ABOUT USED TRANSCEIVERS?

So far we've talked about buying new radios, but there are a large number of hams that prefer used equipment. Buying used will save you money in most cases. Of course, the

The Yaesu FT-857D is a dc-to-daylight mobile/base transceiver selling at less than $700.

The ICOM IC-718 is a low-end HF-only transceiver selling at less than $600.

The Ten-Tec Jupiter is a moderately priced HF-only transceiver.

The Yaesu FTdx-9000 is an example of a high-end HF transceiver. Depending on the model, it sells for more than $10,000.

older the radio, the less compatible it will be with modern technology. Older radios also have a tendency to break down (they have suffered years of wear and tear, after all). When this happens, replacement parts and repair services may be difficult to find.

ROLL YOUR OWN?

Most hams buy their radios factory assembled, but it is possible to *build* your radio as well. QRP enthusiasts, in particular, are fond of building their own transceivers from kits, or even just from diagrams published in books and magazines. Most of QRP transceivers are single-band, CW-only rigs, but there are exceptions.

> *As a rule of thumb, stick with used transceivers that are less than 10 years old if you lack the technical skills to do your own repairs.*

Kit building is fun and educational, and you'll save a considerable amount of money in the process. If you think your technical skills are marginal, however, build your kit with the help of a more knowledgeable ham.

WHEN YOU NEED MORE POWER...

With a few exceptions, HF transceivers produce 100 W output at full power. For most of the operating you'll ever do on the HF bands, this is more than enough, especially if your antenna system is working at peak performance. VHF and UHF transceivers (except for handhelds) usually offer 50 to 100 W maximum output. This is also adequate for most VHF/UHF applications.

But there are times when you simply need more power to make a contact. Maybe the bands are in poor condition, or perhaps you are trying to be heard through a wall of interference. If the output of your transceiver isn't getting the job done and you can't improve your antenna system, the only alternative is to increase your power with an *RF power amplifier*.

Most HF power amps generate between 500 and 1500 W. This is serious power! At VHF and UHF you'll also find amplifiers capable of producing 1500 W. This kind of power must be respected. Lethal voltages are used in these amplifiers; the RF energy itself

The ICOM IC-7000 is a popular wide range (1.8 to 440 MHz) transceiver that can be used in mobile or base applications.

This Elecraft HF transceiver can be built from a kit!

GØBPU built this tiny 80-meter CW QRP transceiver by hand in a single evening.

Shopping for Equipment

If you are looking for new ham equipment, one of your best resources is *QST* magazine, the journal of the American Radio Relay League (ARRL). The magazine is mailed to you each month as part of your ARRL membership (call 860-594-0200, or see the League on the Web at **www.arrl.org**). *QST* meticulously reviews new Amateur Radio equipment in each issue. The magazine also carries advertisements from ham dealers and manufacturers so you can keep up to date with all the new products on the market.

If you have a ham dealer near you, that's a great place to shop. At the dealer's store you can get your hands on the radio and ask questions about it. If you can't get to the store, however, most dealers have Web sites and toll-free telephone numbers. Just look at their advertisements in *QST*.

If you prefer used equipment, you'll find that most of these sales occur online. eBay (**www.ebay.com**) has tons of ham gear for auction every day. Another popular online site is the ARRL's Radios On-Line at **www.arrl.org/RadiosOnline/**.

Dealers often sell used equipment in their stores and on their Web sites. The dealer advantage is that they offer limited used-equipment warranties. This takes much of the worry out of your used-equipment purchase.

It probably goes without saying that you need to be careful when purchasing used equipment. See the equipment and operate it "in person" whenever possible. When shopping eBay auctions, look for sellers that have very good "feedback" ratings, ideally 100% positive. Make sure you read the auction description very carefully. If you have questions, e-mail the seller before you bid.

is sufficient to cause nasty burns to anyone unlucky enough to touch the antenna.

For your first Amateur Radio station, you do *not* need an amplifier. Save your money and wait until you have some experience. You may find that you don't need the extra power after all. If you do, you'll find amplifier reviews in *QST* and on the ARRL Web. As you do your shopping, make sure to check the amplifier specifications concerning their electrical requirements. Amplifiers usually have their own power supplies, but some require 220 Vac, or 120 Vac at higher-than-normal current. You may need to recruit an electrician to wire a dedicated outlet in your station for the amplifier.

Your antenna system needs to be able to handle the increased output. (Antennas *can* melt!) If you have a device in the system such as an antenna tuner, it must be rated for the power as well.

2 The Antenna— The Most Important Part of Your Station

When it comes to getting the most enjoyment out of Amateur Radio, there is nothing more important than your antenna system. You can purchase the best ham transceiver on the planet, but if your antenna system is poor, your investment will largely go to waste.

When we say "antenna system," we're talking about the antenna and everything that connects it to your radio, including the feed line. Let's concentrate on the antennas first.

When you think of an antenna, what sort of image appears in your mind? Do you see a gleaming steel tower, majestically supporting a huge rotating antenna? That's the vision most of us conjure. The radio tower is an ancient icon in our hobby. It symbolizes the art and mystery of radio itself.

> *You can purchase the best ham transceiver on the planet, but if your antenna system is poor, your investment will largely go to waste.*

But ham antennas come in almost every design imaginable. Some are little more than strands of finely tuned wire. Others are thorny javelins of polished metal. Some sit atop towers. Others don't. Some look like monstrous spider webs, while others are modern-art sculptures of aluminum tubing.

ANTENNAS FOR THE HF BANDS

The most powerful antenna for the HF bands or any band is the *directional* antenna, often referred to as the *beam* antenna.

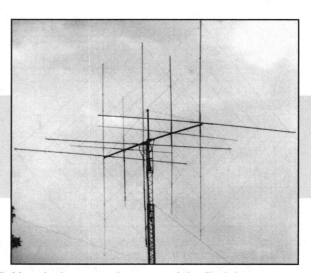

Figure 2.1—The classic HF Yagi antenna (left) and a large quad antenna right. Both beam designs use rotators to turn them in the desired directions.

When hams speak of beam antennas, they usually mean the venerable Yagi and quad designs (see **Figure 2.1**). These antennas focus your signal in a particular direction (like a flashlight). Not only do they concentrate your transmitted signal, they allow you to focus your *receive* pattern as well. For example, if your beam is aimed west you won't hear many signals from the east (off the "back" of the beam).

The problems with beam antenna systems are size and cost. HF beams for the lower bands are *big* antennas. At about 43 feet in width, the longest element of a 40-meter coil-loaded Yagi is wider than the wingspan of a Piper Cherokee airplane.

In terms of cost, a sizeable beam antenna and 75-foot crank-up tower will set you back *at least* $2500. Then add about $500 for the antenna rotator, an electric motor that allows you to turn the antenna by remote control. On top of that, add the cost of cables, contractor fees (to plant the tower in the ground) and so on. In the end, you'll rack up about $5000.

If you have that much cash burning a hole in your pocket, by all means throw it at a beam antenna and tower. The rewards will be tremendous and you'll never regret the investment. Between the signal-concentrating ability of the beam and the height advantage of the tower, you'll have the world at your fingertips. Even a beam antenna mounted on a roof tripod can make your signal an RF juggernaut.

In truth, however, only a minority of hams can afford towers these days. Those who manage to scrape together the necessary funds occasionally find themselves the targets of angry neighbors and hostile town zoning boards. (They don't appreciate the beauty of aluminum spires like we do!)

> **Single-band dipoles are among the easiest antennas to build.**

But do you *need* a beam and a tower to enjoy Amateur Radio? The issue isn't whether they're worthwhile (they are). The question is: Are they absolutely necessary? The answer, thankfully, is *no*.

Single-Band Dipoles

You can enjoy Amateur Radio on the HF bands with nothing more than a copper wire strung between two trees. This is the classic *dipole* antenna. It comes in several varieties, but they all function in essentially the same way.

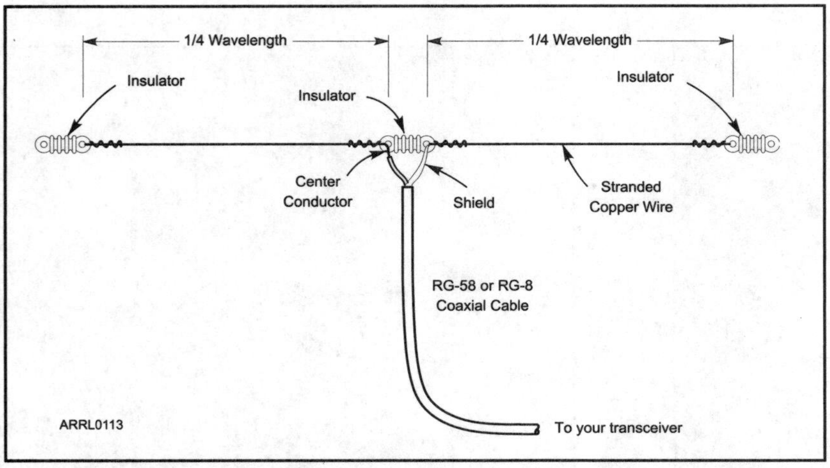

Single-band dipoles are among the easiest antennas to build. All you need is some stranded, noninsulated copper wire and three plastic or ceramic insulators (see **Figure 2.2**). A ½-wavelength dipole is made up of two pieces of wire, each ¼-wavelength long.

Calculating the lengths of the ¼-wavelength wires is simple. Just grab a calculator and perform the following bit of division:

Length (feet) = 468/frequency (MHz)

Actually, you should add about six inches to the results of your calculations. You'll need that length

Figure 2.2—The ½-wavelength dipole is one of the easiest HF antennas to build. All you need is stranded copper wire and three ceramic or plastic insulators. You can use 50-Ω coaxial cable to connect this antenna to your radio, but you'll need to trim the ends to get the best SWR on your favorite frequency range.

There is No Such Thing as a Free Lunch

When you are shopping for antennas, be careful. You'll see all sorts of claims that seem incredible, yet highly attractive. Always follow the old saying, "If it seems too good to be true, it probably isn't."

The No-Radial Vertical

Yes, it is possible to design a vertical antenna that doesn't use radials. But unless the antenna in question is a vertical dipole, you still need a pathway for the RF currents to return to the antenna. This means radials. Without radials, a vertical antenna can still radiate, but not very well. Steer clear of any vertical antenna that claims great performance without radials.

Miniature HF Antennas

The lower the frequency, the bigger the antenna. So how do some manufacturers get away with claiming that their low-band HF antennas will fit in a suitcase? What they have done is use a creative combination of coils and other components to achieve a feed line match at the desired frequency. But a 1:1 SWR does not mean that an antenna is efficient, and efficiency is what really counts. In these tiny HF antennas, most of your precious RF energy is lost as heat. Unless you can't get on the air any other way, choose a full size antenna instead.

"100% Efficient!"

An antenna that is 100% efficient radiates all of its RF energy without a single microwatt of loss. However, unless the antenna is made out of exotic as-yet-undiscovered materials, 100% efficiency is impossible. If you see an advertisement that makes this claim, turn the page quickly.

Wild Gain Claims

If you glance through the pages of *QST* magazine, you'll notice that very few antenna advertisements include gain figures. Antenna gain is measured in decibels (dB) and describes how powerfully the antenna directs your RF energy. The reason for the lack of gain figures is that *QST* magazine has a strict policy: If you claim a gain figure for your antenna, you must be able to prove it.

The problem is that anyone can build an antenna and make all sorts of wild claims about its performance. Unless they've gone to the trouble of having it analyzed using antenna modeling software, or tested on a laboratory grade antenna test range, there is no way to know if what they are saying is true.

That's why *QST* requires proof; most other ham magazines don't. When you read an advertisement or an article where antenna gain figures are tossed around, be careful. Ask where those decibel figures came from!

"Less than 1.5:1 SWR on every band!"

An antenna that can give you an SWR of 1.5:1 on any band is either grossly inefficient, or the advertisement isn't telling you that the low SWR is only present through a *limited portion* of every band. If the manufacturer is claiming a low SWR on any frequency within any band, take your money elsewhere!

margin to trim and tune for the lowest SWR. See the sidebar, "The Importance of SWR."

Split this length of wire into two equal halves and join them in the center with an insulator, then place insulators at both ends. Solder the center conductor of your coaxial cable feed line to one side of the center insulator. (It doesn't matter which side.) Solder the shield braid of your cable to the other side. Connect ropes, nylon string or whatever to the end insulators and haul your antenna skyward. Get it as high as you can and as straight as possible. Don't hesitate to bend your dipole if that's what it takes to make it fit.

Once your dipole is safely airborne, fire up your transmitter and check the SWR at many points throughout the band. (It helps if you can plot the results on graph paper.) If you see that the SWR is getting *lower* as you move lower in frequency, your antenna is too long. Trim a couple of inches from each end and try again. On the other hand, if you see that the SWR is getting *higher* as you go lower in frequency, your antenna is too short. You'll need to *add* wire to both ends and make another series of measurements.

When you've finished trimming your dipole, you'll probably end up with an SWR of 1.5:1 or less at the center frequency, rising to 2:1 or somewhat higher at either end of the band. Don't expect a 1:1 SWR across the entire band. By carefully trimming the antenna you can move the low-SWR portion to cover your favorite frequencies.

Trap Dipoles and Parallel Dipoles

For multiband applications, you'll often find the *trap* dipole (**Figure 2.3**) and the *parallel* dipole (**Figure 2.4**). Traps are tuned circuits that act somewhat like automatically switched inductors or capacitors, adding or subtracting from the length of the antenna according to the frequency of your signal. The parallel dipole uses a different approach. In the parallel design, several dipoles are joined together in the center and fed with the same cable. The dipole that radiates the RF is the one that presents an impedance that most closely matches the cable (50 Ω). That matching impedance will change according to the frequency of the signal. One

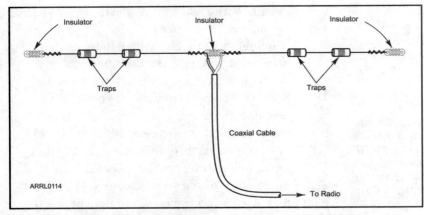

Figure 2.3—A trap dipole uses tuned circuits known as "traps" to electrically shorten or lengthen the antenna. This allows a single antenna to operate on several bands.

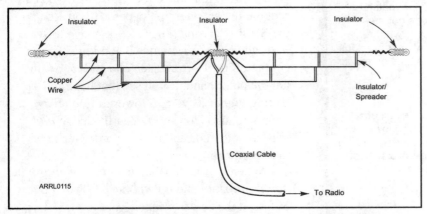

Figure 2.4—The parallel dipole is really several individual dipole antennas connected to the same center insulator. The dipoles tend to interact with one another, so trimming them for the best SWR on each band can be quite difficult.

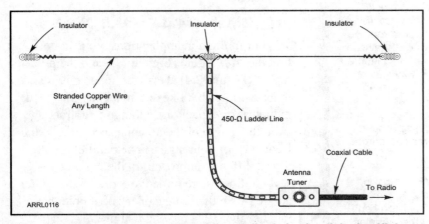

Figure 2.5—The random length multiband dipole is simplicity itself! Just make each leg as long as possible, keeping the lengths equal. Then, feed the antenna in the center with ladder line and use a balanced antenna tuner to tune the antenna for the best SWR as you change bands.

dipole will offer a 50-Ω match on, say, 40 meters, while another provides the best match on 20 meters.

Obviously, these designs are somewhat more complicated than monoband dipoles, although many hams *do* choose to build their own. (See *The ARRL Antenna Book* for construction details.) If you don't have time or desire to tackle a trap or parallel dipole, you'll discover that many *QST* magazine advertisers sell prebuilt models.

Random-Length Multiband Dipoles

You can also enjoy multiband performance *without* traps, coils or other schemes. Simply cut two equal lengths of stranded copper wire. These are going to be the two halves of your dipole antenna. Don't worry about the total length of the antenna. Just make it as long as possible. You won't be trimming or adding wire to this dipole.

Feed the dipole in the center with 450-Ω *ladder line* (available from most ham dealers), and buy an *antenna tuner* with a *balanced output* (see **Figure 2.5**). Feed the ladder line into your house, taking care to keep it from coming in contact with metal, and connect it to your tuner. Use regular coaxial cable between the antenna tuner and your radio.

You can make this antenna yourself, or buy it premade if you're short on time. A 130-foot dipole of this type should be usable on almost every HF band. Shorter versions will also work, but you may not be able to load them on every band.

Ladder line offers extremely low RF loss on HF frequencies,

even when the SWR is relatively high. Just apply a signal at a low power level to the tuner and adjust the tuner controls until you achieve the lowest SWR reading. (Anything below 2:1 is fine.) You'll probably find that you need to readjust the tuner when you change frequencies. (You'll *definitely* need to readjust it when you change bands.)

You may discover that you cannot achieve an acceptable SWR on some bands, no matter how much you adjust the tuner. Even so, this antenna is almost guaranteed to work well on several bands, despite the need to retune.

So why doesn't everyone use the ladder line approach? The reason has much to do with convenience. Ladder line isn't as easy to install as coax. You must keep it clear of large pieces of metal (a few inches at least). Unlike coax, you can't bend and shape ladder line to accommodate your installation. And ladder line doesn't tolerate repeated flexing as well as coaxial cable. After a year or two of playing tug o' war with the wind, ladder line will often break.

Besides, many hams don't relish the idea of fiddling with an antenna tuner every time they change bands or frequencies. They enjoy the luxury of turning on the radio and jumping right on the air—without squinting at an antenna tuner's SWR meter and twisting several knobs.

Even with all the hassles, you can't beat a ladder-line fed dipole when it comes to sheer lack of complexity. Wire antennas fed with coaxial cable must be carefully trimmed to render the lowest SWR on each operating band. With a ladder line dipole, no pruning is necessary. You don't even care how long it is. Simply throw it up in the air and let the tuner worry about providing a low SWR for the transceiver.

> *...you can't beat a ladder-line fed dipole when it comes to sheer lack of complexity.*

Whichever dipole you finally choose, install it as high as possible. If a horizontal dipole is too close to the ground, the lion's share of your signal will be going skyward at a steep angle. Without wading chest deep into propagation analysis, the bottom line is that a high radiation angle is generally not good for long-distance communication. Forty to 70 feet is generally considered the ideal height range, but don't lose sleep if you fall short. Raise the antenna as high as you can and change the subject when you're asked about it. You'll still make lots of contacts.

Verticals

The vertical is a popular antenna among hams who lack the space for a beam or long wire antennas. In an electrical sense, a vertical is a dipole with half of its length buried in the ground or "mirrored" in its ground system. Verticals are commonly installed at ground level, although you can also place a vertical on the roof of a building.

At first glance, a vertical looks like little more than a metal pole jutting skyward. A single-band vertical may be exactly that! However, if you look closer you'll find a network of wires snaking away in all directions from the base of the antenna. In many instances, the wires are buried a few inches beneath the soil. These are the vertical's *radials*. They provide the essential ground connection that creates the "other half" of the antenna. Multiband verticals use several traps or similar circuits to electrically change the length of the antenna according to the frequency of the transmitted signal. (The traps are in the vertical elements, not the radials.)

Vertical antennas take little horizontal space, but they can be quite tall. Most are at least ¼-wavelength long at the lowest frequency. To put this in perspective, an 80-meter

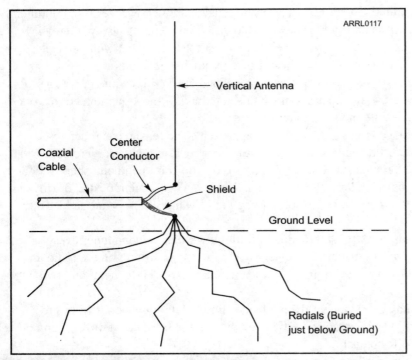

Figure 2.6—A vertical antenna is little more than a vertical pipe or wire, typically ¼-wavelength long at the desired frequency. Think of it as a dipole with the "other half" of the antenna formed by the radials. As you can see in this drawing, the radials don't need to be arranged in straight lines. They don't even have to be buried if you're sure no one will trip over them. Some vertical antennas use an elaborate system of traps and other techniques to operate on more than one band.

full-sized vertical can be over 60 feet tall! Then there is the space required by all those radial wires. You don't have to run the radials in straight lines (see **Figure 2.6**). In fact, you don't even have to run them underground. But you *do* need to install as many radials as possible for each band on which the antenna operates. Depending on the type of soil in your area, you may get away with a dozen radials, or you may have to install as many as 100.

Contemplate spending several days on your hands and knees pushing radial wires beneath the sod. It isn't a pretty picture, is it? That's why several antenna manufacturers developed verticals that do not use radials at all. The most efficient of these verticals are actually *vertical dipoles*. Yes, they are dipole antennas stood on end! There is no reason why this cannot be done. In fact, a vertical dipole can work quite well.

So how does a traditional vertical antenna stack up against a traditional horizontal dipole when it comes to performance? If you have a generous radial system, the vertical can do at least as well as a dipole in many circumstances. Some claim that the vertical has a special advantage for DXing because it sends the RF away at a low angle to the horizon. Low radiation angles often mean longer paths as the signal bends through the ionosphere.

Without a decent radial system, however, the vertical is a poor cousin to the dipole. The old joke, "A vertical radiates equally

> *If you have a generous radial system, the vertical can do at least as well as a dipole in many circumstances.*

poorly in all directions," often applies when the ground connection is lacking, such as when the soil conductivity is poor. If you can't lay down a spider web of radials, dipoles are often better choices.

Random Wires

A random wire is exactly that—a piece of wire that's as long as you can possibly make it. One end of the wire attaches to a tree, pole or other support, preferably at a high point. The other end connects to the random-wire connector on a suitable antenna tuner (**Figure 2.7**). You apply a little RF and adjust the antenna tuner to achieve the lowest SWR. That's about all there is to it.

Random-wire antennas seem incredibly simple, don't they? The only catch is that your antenna tuner may not be able to find a match on every band. The shorter the wire,

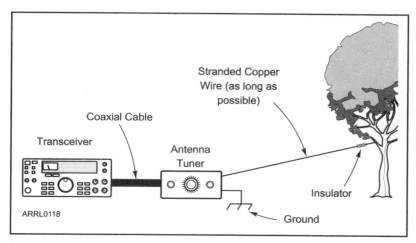

Figure 2.7—A random end-fed wire antenna is about as simple as it gets, but you need an antenna tuner to provide a match to your radio.

the fewer bands you'll be able to use. And did you notice that the random wire connects directly to your antenna tuner? That's right. You're bringing the radiating portion of the antenna right into the room with you. If you're running in the neighborhood of 100 W, you could find that your surroundings have become rather hot—*RF* hot, that is! We're talking about painful "bites" from the metallic portions of your radio, perhaps even a burning sensation when you come in contact with the rig or anything attached to it.

Random wires are fine for low-power operating, however, especially in situations where you can't set up a vertical, dipole or other outside antenna. And you may be able to get away with higher power levels if your antenna tuner is connected to a good Earth ground. (A random-wire antenna needs a good ground regardless of how much power you're running.) If your radio room is in the basement or on the first floor, you may be able to use a cold water pipe or utility ground. On higher floors you'll need a *counterpoise*.

A counterpoise is simply a long, insulated wire that attaches to the ground connection on your antenna tuner. The best counterpoise is ¼-wavelength at the lowest frequency you intend to use. That's a lot of wire at, say, 3.5 MHz, but you can loop the wire around the room and hide it from view. The counterpoise acts as the other "terminal" of your antenna system, effectively balancing it from an electrical standpoint.

Indoor Antennas

So you say that you can't put up an outdoor HF antenna of any kind? There's hope for you yet. Antennas generally perform best when they're out in the clear, but there is no law that says you can't use an outdoor antenna *indoors*.

If you have some sort of attic in your home, apartment or condo, you're in luck. Attics are great locations for indoor antennas. For example, you can install a wire dipole in almost any attic space. Don't worry if you lack the room to run the dipole in a straight line. Bend the wires as much as necessary to make the dipole fit into the available space.

Of course, this unorthodox installation will probably require you to spend some time trimming and tweaking the length of the antenna to achieve the lowest SWR (anything below 2:1 is fine). Not only will the antenna behave oddly because of the folding, it will probably interact with nearby electrical wiring.

This 40-meter loop antenna found a home in the attic of OK0EU.

The Importance of SWR

SWR stands for Standing Wave Ratio. A discussion of SWR can become extremely technical, but it doesn't need to be. Simply put, SWR is an indication of how well your antenna system is matched to your radio.

Most transceivers you'll encounter are designed to expect a 50-Ω (ohm) impedance at their outputs. When the antenna system impedance is 50 Ω, the transceiver can safely transfer its RF output power. But when the impedance is something other than 50 Ω, a strange thing happens. The RF energy from the transceiver still travels to the antenna, but a portion of it is reflected back to the radio as though the antenna was a mirror. This energy zips back to the transceiver where it is reflected back toward the antenna!

Back and forth, back and forth, the RF energy endlessly travels. By doing so, "standing waves" of energy are created in the feed line. Think of a motor that constantly agitates a pool of water. The waves hit the edges of the pool and bounce back, over and over, until it looks like the waves are standing still in the water. These are the standing waves.

An SWR of 1:1 means that nearly all of the RF power generated by the transceiver is making it to the antenna without being reflected. This is the ideal situation. As the SWR increases, power is being reflected and is turning into heat (and being lost) within the feed line. This reflected energy is also creates substantial voltages at the transceiver output, which can cause damage.

Modern transceivers are designed to reduce their output power when the SWR rises above about 2:1. So, you need to keep the SWR at your transceiver as low as possible—preferably less than 2:1. A high SWR can not only damage your radio, it is also wasteful since so much energy— transmit *and receive* energy—is being lost.

You can determine the SWR at your transceiver with a device known as an SWR meter. This bit of electronic wizardry measures the Standing Wave Ratio (SWR) at the point where you install it in the feed line. Most of the time the meter is installed at your transceiver. In fact, many transceivers have SWR meters already built-in.

If your radio doesn't include an SWR meter, buy one. You won't regret the purchase. An SWR meter can tell you if your radio is generating the power it should, and if your antenna system is properly matched. Most hams leave their SWR meters in the line at all times, usually near the transceiver. This way, if a problem occurs, they'll know immediately.

For instance, if a coax connection fails at your antenna, you'll know because you will see the SWR suddenly rise when you are transmitting. A loose connection often manifests itself as an SWR that jumps wildly up and down as you transmit.

SWR meters are not foolproof. You can have an antenna with a high SWR at the point when the feed line connects to it. Losses in the feed

An SWR meter is one of the best pieces of test equipment that you can buy for your station. SWR meters come in several varieties. This Daiwa model uses a "cross needle" movement. One needle shows forward power, the other needle indicates reflected power and where they cross is the SWR.

line between the antenna and the station, however, can effectively "fool" the meter into displaying an SWR that is lower than it really is.

When shopping for an SWR meter, look for a model that measures power as well as SWR. These are sometimes called SWR/wattmeters. Also make sure that the meter covers the frequency range you require. An SWR meter designed for the HF bands will display very inaccurate readings when used at VHF and UHF.

Finally, beware of "bargain" meters. Accurate, well-made test equipment always costs a little more. This is one purchase where you definitely don't want to be penny wise and pound foolish.

Ladder-line fed dipoles are ideal for attic use—assuming that you can route the ladder line to your radio without too much metal contact. In the case of the ladder-line dipole, just make it as long as possible and stuff it into your attic any way you can. Let your antenna tuner worry about getting the best SWR out of this system.

The same dipoles and loops that you use in your attic can also be used in any other room in your home. The same techniques apply. Keep the antenna as high off the floor as possible. (As with most antennas, the more height, the better.) For indoor operating, however, use low output power. You'll avoid RF "bites" as well as interference to VCRs, TVs and so on. Many hams have been successful operating indoor antennas with just a few watts output.

ANTENNAS FOR VHF AND UHF

Antennas for the VHF and UHF bands are similar in many ways to HF antennas. The main differences are that VHF/UHF antennas are smaller and the losses caused by poor feed lines and elevated SWRs (or both) are more critical.

Omnidirectional Antennas

This type of VHF antenna transmits and receives in all directions at once (the same is true of the dipoles, loops and vertical antennas for HF use). All commonly used mobile antennas are omnidirectional. This makes sense because it is impractical to stop and point your car in the direction of the station you want to contact. Instead, the omnidirectional mobile antenna blasts your signal in all directions so that you'll stand a decent chance of communicating no matter where you are driving.

Omnidirectional antennas are also found in base stations where the goal is to transmit and receive from any direction with minimal hassle and expense. Common omnidirectional antenna designs for base stations include *ground planes* (see **Figure 2.8**), *loops* and *J-poles*, but there are others.

An omnidirectional antenna spreads your signal over a broad area, depending on how high you install it. Height is critical to the performance of all antennas at VHF and UHF frequencies. Higher is always better, whether that means putting the antenna on a flagpole, tower or a rooftop. If you are fortunate enough to operate from the summit of a hill or mountain where Mother Earth provides the altitude, that works, too.

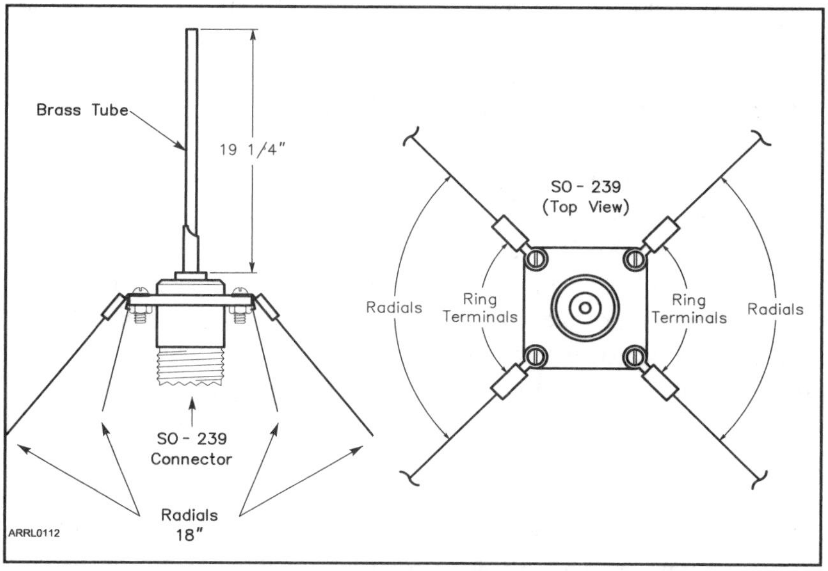

If the advantage of an omni is that it radiates in all directions, **that** can be its *disadvantage* as well. **An** omnidirectional antenna can't focus your reception or transmission. Once you put it in place, what you get is…well…what you get. There is little you can do to change it. If the station you're talking to is west

Figure 2.8—A VHF or UHF ground plane omnidirectional antenna is absurdly simple to build. The ground plane shown here is for the 2-meter band.

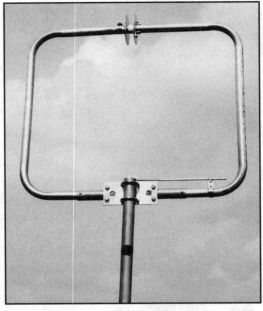

A 6-meter loop antenna known as a "squalo."

of your location, for example, all the power you are sending north, south and east is wasted. You will also receive signals—possibly interfering signals—from the same useless directions.

Directional Antennas

As the name implies, directional "beam" antennas focus your power and reception in a single direction. Just like HF antennas, directional VHF designs work by canceling the energy that radiates toward the back of the antenna and reinforcing the energy going toward the front. The result is a beam of RF power (and concentrated receive sensitivity) not unlike a searchlight or a magnifying glass.

Directional antennas are ideal at VHF and UHF when you want maximum distance and minimum interference. They are almost mandatory for VHF DX work and satellite operating. Directional antennas also help tremendously on VHF FM when you're trying to communicate with a distant station. Common directional antenna designs include the *Yagi, quad* and *Moxon. Parabolic dish* antennas—the kind you've likely seen for satellite TV reception—are also directional antennas.

So what is the downside?

Directional antennas tend to be more complex and difficult to assemble. They can also be quite large in some configurations. For instance, a highly directional Yagi antenna for the 6-meter band, a model with 11 sections known as *elements*, can include a boom assembly that's nearly 70 feet in length.

And what happens if your antenna is pointing north and the station you want to talk to is south? Unless you can turn your antenna, communication will be difficult or impossible. This is where the *antenna rotator* comes into play, just as it did for HF beam antennas. You may recall that a rotator is an electric motor that you install below your directional antenna. Its job is to turn your antenna to the direction you require.

Rotators add to the cost and complexity of a directional antenna

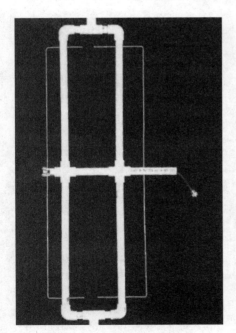

This interesting VHF beam antenna is called a Moxon after Les Moxon, G6XN, who came up with the original design.

system. A light duty rotator can cost about $100. If you need a heavy duty rotator to turn a bigger antenna (or more than one antenna), the cost can reach $500 or more. In addition to the hassle of stringing your feed line from the antenna back to your radio, you must also string a cable for the rotator. More wires equal more work, although the reward can be considerable!

Antenna Polarity

Strange as it may sound, antennas have polarities. Not plus and minus like a batteries, but horizontal, vertical or circular. Antenna polarity is unimportant at HF, but it is critical at VHF and UHF. On the VHF/UHF bands, matching the polarity of the station you are talking to can make a substantial difference in signal strength. If you install a directional antenna with its elements perpendicular to the ground, it will have vertical signal polarity. If you mount it parallel to the ground, its polarity will be horizontal.

Omnidirectional antennas can also have horizontal or vertical polarity. Most omni designs are vertically polarized, but certain antennas such as loops are often designed for horizontal polarity.

Circular polarization is primarily used for satellite antennas. With this design, the signal "rotates" through vertical and horizontal as it travels. This is helpful when you are trying to communicate with a spacecraft that is constantly changing its position (and polarity) relative to your station.

So which polarity should you use? In Amateur Radio, the unwritten rule is that vertical polarization is used for VHF/UHF FM work. VHF stations operating mostly SSB or CW use horizontal polarization.

This low-cost TV antenna rotator is adequate to turn small beam antennas, say for the 10 or 15 meter HF bands, or for VHF and UHF.

FEED LINES

Regardless of whether you are operating at HF, VHF or UHF, the quality of your feed line is critical to your station. The feed line (also called the *transmission line*) is the RF power conduit between your radio and your antenna. All the energy you generate travels to the antenna through the feed line. By the same token, all the signals picked up by your antenna must reach your radio through the same feed line.

The problem with any feed line is that it isn't perfect—it always loses a certain amount of energy. To complicate matters, all feed lines are not created equal. The amount of loss at any frequency will vary considerably from one type of feed line to another.

The most common type of feed line is *coaxial cable*, or simply *coax*. It is called coaxial because there are two circular conductors positioned "co-axially" (on the same axis), one inside the other. The inner conductor is usually called the "center conductor." It is surrounded by a solid or multistranded outer conductor commonly called a "shield." The shield is usually surrounded by an insulating

> *The problem with any feed line is that it isn't perfect—it always loses a certain amount of energy.*

Antenna Tuners

Consider an antenna tuner as an adjustable impedance matcher between your antenna system and your transceiver. An antenna tuner doesn't really "tune" the antenna in a literal sense.

Antenna tuners come in all sizes according to how much RF power they can handle and the designs of their matching circuits. Prices range from about $80 to well over $1000. If you're using the typical 100-W transceiver, you should be able to find an adequate tuner in the $120 to $200 price range.

If you intend to use an antenna fed with ladder line, or a random-wire antenna, you *must* have an antenna tuner. For coaxial-fed antennas, however, tuners are optional. If the SWR at your operating frequency is less than 2:1 on a coax-fed antenna, you don't need a tuner. (Besides, some HF transceivers feature built-in tuners.)

Remote Automatic Antenna Tuners

An interesting variation on the antenna tuner idea is the remote automatic antenna tuner. Remote antenna tuners are usually housed in nondescript weatherproof enclosures. They are designed for installation *at the antenna*, not indoors at your station. When you transmit, the tuner automatically tunes for the lowest SWR. Since the tuner is sitting right at the antenna, you don't have to worry about RF loss in your feed line—the SWR between your transceiver and the tuner will always be as low as possible.

Remote automatic antenna tuners can be expensive, typically $300 or more. Most of these tuners also require you to run a separate dc power line between your station power supply and the tuner. Despite the cost and installation hassle, remote automatic antenna tuners are extremely convenient and versatile. For instance, you can place a remote tuner at the center of a single-band dipole antenna and get *multiband* use from the same antenna!

Figure 1A—A manual antenna tuner.

Figure 2A—An SGC remote automatic antenna tuner.

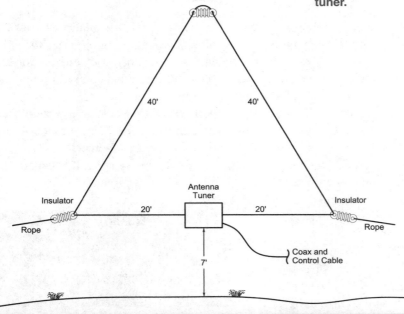

Figure 3A—In this example, a remote automatic antenna tuner is part of a vertical three-sided loop antenna known as a delta loop. Whenever the tuner senses RF power from the transceiver, it tunes the loop for the lowest SWR.

Table 2-1

A few common coaxial cables and their loss ratings at 3.5, 144 and 440 MHz.

	Loss (dB per 100 feet)		
Cable Type	3.5 MHz	144 MHz	440 MHz
LMR-600	0.13	0.95	1.70
LMR-400	0.22	1.50	2.75
9913	0.21	1.60	2.80
RG-213	0.37	2.80	5.20
RG-58	1.00	4.00	7.00

plastic jacket. There is also insulating material between the center conductor and the shield. This material can be hard plastic, foam plastic or even air.

A popular type of feed line for HF use is ladder line. In fact, at HF frequencies it is the most common feed line for random-length dipoles and other antenna designs. Ladder line consists of nothing more than two wires in parallel separated by insulating material.

When rating feed lines for loss, we use "decibels (dB) per 100 feet." If you're not familiar with the decibel, don't worry. Just remember that the higher the decibel number, the greater the loss. In **Table 2-1** you'll find a list of common types of coaxial cable and their loss figures at 80 meters (3.5 MHz) 2 meters (144 MHz) and 70 cm (440 MHz).

Feed lines also have a characteristic *impedance* value measured in *ohms* (Ω). Coaxial cable commonly used for Amateur Radio has an impedance of 50 Ω while ladder line impedances can vary from 300 to 600 Ω. Amateur Radio transceivers are designed to work with an impedance of 50 Ω, so you must use 50 Ω coax, or find a way to convert the 300 to 600 Ω impedance of ladder line to 50 Ω. If your radio "sees" anything other than 50 Ω, it will reduce its output to protect itself from the possible damage that can result in a high SWR condition.

If you are using an antenna that is designed to deliver a 50-Ω impedance, it is best to use a coaxial feed line to provide a 50-Ω antenna system impedance for your transceiver. Even these 50 Ω antennas can be a little "off" at times, so you may need to tune them by physically cutting or adjusting the antenna to the correct length, as we discussed earlier, or by adjusting a matching section at the antenna.

Coaxial cable. Note how the solid center wire is surrounded by a plastic insulating material and a braided wire shield. The cable on the top includes a solid metal shield as well.

The other approach is to use a device called an antenna tuner to transform the impedance of the antenna system to 50 Ω for your radio without physically adjusting the antenna at all. An antenna tuner is a kind of adjustable impedance transformer. Some tuners operate manually; you twist the knobs until the SWR meter shows a 1:1 SWR, or something reasonably close to it. Other tuners are automatic and do all the adjustments for you.

Taking the antenna tuner approach is not a good idea when you are using coaxial cable under high (greater than 3:1) SWR conditions. The tuner may provide the 50 Ω match to your radio, but the mismatch and high SWR *still exists between the antenna tuner and the antenna!* This translates to high losses in the coaxial cable.

On the other hand, using an antenna tuner with ladder line is a good way to go—at least for HF work. At HF frequencies, the loss in ladder line is so low, you can still see good results even when the SWR is horrendous. The antenna tuner provides the 50 Ω match to your radio and you really don't care what the SWR is between the tuner and the antenna.

So which type of feed line should

You want the feed line that has the lowest loss at the highest frequency you want to operate.

HF Mobile Antennas

Most HF mobile antennas are compromises. They sacrifice efficiency for reduced size. Even so, they work well enough to allow you to enjoy conversations when you're on the road.

Some mobile antennas use a fixed-frequency design. They are adjusted for the lowest SWR at a particular frequency, then left in place. When you change bands, you must change antennas.

Other mobile antennas are equipped with small electric motors that adjust their lengths and effectively retune the antennas whenever you change frequencies. These so-called "screwdriver" antennas (so named because of the electric screwdriver motors they use) can be expensive, but they make up for the cost with extraordinary convenience.

All HF mobile antennas require good ground connections to the metal chassis of the vehicle to perform well. They also have to be well connected to the car so that they can survive wind and vibration.

The screwdriver-type HF mobile antenna shown hear almost dwarfs the automobile!

you use at your station? Fortunately, the answer is simple: You want the feed line that has the lowest loss at the highest frequency you want to operate.

As you probably guessed, low-loss feed lines are more expensive. Some of the low-loss feed lines are also rigid and hard to work with (they don't bend easily). A little planning and common sense goes a long way when it comes to selecting feed line.

In a mobile installation, you can use an inexpensive feed line such as RG-58 because you're only using a short length. As long as the SWR is low, the loss will be acceptable.

However, if you have an antenna that is 100 feet from your radio and you are operating at, say, 440 MHz, RG-58 would be an extraordinarily bad choice! For this installation you'll need to invest in something much better—probably LMR-400 or Belden 9913 (see Table 2-1).

For base stations in particular, always buy the lowest-loss coax you can afford. Since you'll probably be using your feed line for several years or longer, you want something that can support your changing interests. For instance, 100 feet of LMR-400 is overkill quality for a station that only operates on the 40-meter band. But if you someday want to switch to 440 MHz, you'll be glad that you already have a low-loss feed line in place!

The Care and Feeding of Feed Lines

In order to prevent excessive losses, feed lines must be protected. Coaxial cable depends on the integrity of its outer coating—the jacket—to keep water out. Nicks, cuts, and scrapes can all breach the jacket. Water in coaxial cable degrades the effectiveness of the braided shield and dramatically increases losses. Coax cannot be bent sharply, lest the center conductor be forced gradually through the soft center insulation, eventually causing a short. Prolonged exposure to the ultraviolet (UV) in sunlight will also cause the plastic in the jacket to degrade, causing small cracks. Coax that has been exposed to the weather for a long time often has losses much higher than that of new cable.

Open-wire feed lines also need care. These feed lines often are constructed using solid wire. Prolonged flexing in the wind will eventually crack and break the conductors if no strain relief is provided. Moss, vines, or lichen growing on the cable will also increase loss. Tree limbs rubbing on the line will eventually break it. Protect splices from weather damage with good-quality electrical tape or a paint-on coating.

Connectors for coaxial cable ("coax connectors") are required to make connections to radios, accessory equipment, and most antennas. "Pigtail" style connections, where the braid and center conductor are separated and attached to screw terminals are generally unsuitable at frequencies above HF. Pigtails also expose the cable to water.

3 Propagation—The Science of How Signals Travel

With contributions from Ward Silver, NØAX

Understanding how signals travel—how they propagate—is essential to getting maximum enjoyment from Amateur Radio. Think of yourself as an angler fishing for signals. To catch the most "fish," you need to know how they are likely to behave under various conditions.

For example, cold, muddy water is a lake fisherman's nightmare. Under these conditions he knows that he has to cast his bait close to "cover" (plants, etc) to have any hope of success. That's because fish lurk near cover when water is muddy.

> *Propagation is a strange and quirky thing. It is full of surprises and genuine mystery.*

By the same token, high noon is a poor time to look for DX contacts on the 80-meter band. The savvy DX angler knows that he was to wait for nightfall, or go up to 20, 17, 15 or 10 meters.

Propagation is a strange and quirky thing. It is full of surprises and genuine mystery. The 6-meter band, for instance, tends to abruptly "open" in the spring and in mid-winter for contacts spanning thousands of miles. This type of propagation is known as sporadic E. What causes sporadic E? Believe it or not, no one is quite sure. Scientists have a handful of theories, but none have been proven thus far.

Don't worry, though. You don't need to take a course in physics to grasp the wonders of radio propagation. Let's just do a quick review of the fundamentals.

MOVING WAVES

Radio waves spread out from an antenna in straight lines unless reflected or diffracted along the way, just like light. Of course, light waves are just a very, very, very high frequency form of radio waves! And like light, the strength of a radio wave decreases as it travels farther from the transmitting antenna. Eventually the

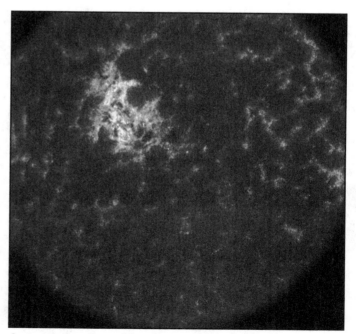

A stormy Sun has a big effect on the HF bands!

Notice this summertime cool front stretching from Michigan down through Illinois and beyond, as well as the high-pressure area in the southeast. If you are a VHF/UHF DX hunter, these conditions can bring band openings!

wave becomes too weak to be received because it has either spread out too much or something along its path absorbed or scattered it. The distance over which a radio transmission can be received is called range.

If the transmitting and receiving antennas are within direct sight of each other, that's line-of-sight propagation. Most propagation at VHF and higher frequencies is line-of-sight. Increasing antenna height or transmitter power also increases the range of line-of-sight propagation. Radio waves at HF and lower frequencies can also travel along the surface of the earth as ground wave propagation.

Radio waves can be reflected by any sudden change in the media through which they are traveling, such as a building, hill, or even weather-related changes in the atmosphere. Obstructions such as buildings and hills create radio shadows, especially at VHF and UHF frequencies. Radio waves can also be diffracted or bent as they travel past edges of buildings or hills. Radio signals arriving at a receiver after taking different paths from the transmitter can interfere with each other if they are out of phase, even canceling completely! This phenomenon is known as multi-path. Signals from a station moving through an area where multi-path is present have a characteristic rapid variation in strength known as mobile flutter or picket-fencing.

Propagation above VHF frequencies assisted by atmospheric phenomena such as weather fronts or temperature inversions is called tropospheric propagation or just "tropo." Radio signals are also reflected by conducting things in the atmosphere, such as planes, meteor trails and even the polar auroras. All of these can reflect signals over hundreds or thousands of miles and are regularly used by amateurs to make short contacts that would otherwise be impossible by line-of-sight propagation. Yet there is one more conductive thing floating around "up there" that hams use every day to communicate around the world.

The Ionosphere

Above the lower atmosphere where the air is relatively dense and below outer space where there isn't any air at all lies the ionosphere. In this region from 30 to 260 miles above the earth, atoms of oxygen and nitrogen gas are exposed to the intense and energetic ultra-violet (UV) rays of the sun. These rays have enough energy to create positively charged ions from the gas atoms by knocking loose some of their negatively charged electrons. The resulting ions and electrons create a weakly conducting region high above the earth.

Above the lower atmosphere where the air is relatively dense and below outer space where there isn't any air at all lies the ionosphere.

Because of the way the atmosphere is structured, the ionosphere forms as layers shown in **Figure 3.1**, called the D, E, F1 and F2 layers, with the D-layer being the lowest. Depending on whether it is night or day and the intensity of the solar radiation these layers can diffract (E, F1, and F2 layers) or absorb (D and E layers) radio waves. The ability of the ionosphere to diffract or bend radio waves also depends on the frequency of the radio wave. Higher frequency waves are bent less than those of lower frequencies. At VHF and higher frequencies the waves usually pass through the ionosphere with only a little bending and are lost to space.

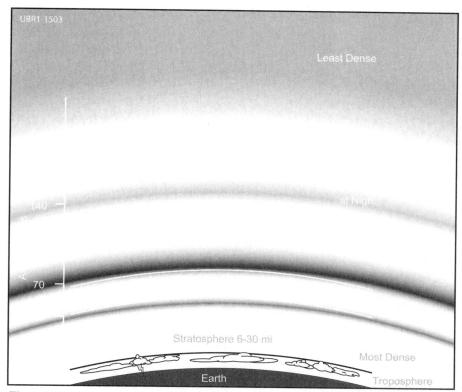

Figure 3.1—The ionosphere forms layers called the D, E, F1 and F2 layers, with the D-layer being the lowest.

Radio waves at HF (and sometimes VHF) can be completely bent back towards the earth by the ionosphere's F-layers as if they were reflected. This is called sky wave propagation or skip. Since the earth's surface is also conductive, it can reflect radio waves, too. This means that a radio wave can be reflected between the ionosphere and the earth multiple times. Each reflection from the ionosphere is called a hop and allows radio waves to be received hundreds or thousands of miles away. This is the most common way for hams to make long-distance contacts. The highest frequency signal that can be reflected back to a point on the earth is the Maximum Useable Frequency (MUF) between the transmitter and receiver. The lowest frequency that can travel between those points without being absorbed is the Lowest Useable Frequency (LUF).

When sky-wave propagation on an amateur band is possible between two points, the band is said to be open. If not, the band is closed. Because the ionosphere depends on solar radiation to form, areas in daylight have a different ionosphere above them than do those in nighttime areas. That means radio propagation may be supported in some directions, but not others, opening and closing to different locations as the earth rotates and the seasons change. This makes pursing long-distance contacts very interesting!

VHF and UHF enthusiasts also experience exciting ionospheric propagation. When solar radiation becomes sufficiently intense, such as during the peak of the 11-year sunspot cycle, the F layers of the ionosphere can bend even VHF signals back to earth. When those ham bands open, they support long-distance communication not possible under usual conditions. In addition, at all points in the solar cycle, patches of the ionosphere's E-layer can become sufficiently ionized to reflect VHF and UHF signals back to earth. This is the sporadic-E (sometimes shown as E_s) propagation mentioned earlier.

A TOUR OF THE MF/HF BANDS

There are nine groups of frequencies—or bands—in the HF/MF region where hams are allowed to operate. Each has its own flavor.

160 Meters: 1.8 to 2 MHz

This is the "basement" of Amateur Radio, the lowest band that we can use. It's almost unusable during daylight hours, but nighttime signals on 160 can travel hundreds or even thousands of miles. Noise levels are high, especially during the summer months when thunderstorms march across the landscape. This makes 160 meters primarily a winter band. You'll find a mix of CW and SSB activity on this band, although most of the long-distance work (DX) takes place on CW.

Because of the large antennas required for 160 operating, it isn't a band for everyone. As a result, crowding is at a minimum. This band can be exotic, almost to the point of being spooky. Signals from halfway around the world can suddenly rise out of the noise like ghosts. If you have the real estate to set up a 160-meter antenna, this band will keep your attention on those cold winter evenings!

80 Meters: 3.5 to 4 MHz

Like 160 meters, 80 meters is considered a nighttime band. Even so, 80 is also good for daytime communication out to a few hundred miles.

Depending on atmospheric conditions, 80 meters can offer worldwide communication at night. Even under mediocre conditions, contacts between the East Coast of the US, for example, and Europe are common. Like 160, this band also suffers from high noise levels, so it is definitely at its best during the winter.

The voice portion of 80 meters is inhabited primarily by SSB (single sideband) operators, although you'll hear a couple of AM signals here and there. When the band is in decent shape, the voice portion can become extremely crowded. This is particularly true between 3.85 and 4 MHz.

You may also find wall-to-wall signals in the CW segment. Chasing international contacts on 80-meter CW is a favorite pastime. Even digital modes such as RTTY and PSK31 can be found on 80 meters. Look for these signals around 3.580 MHz after sunset.

60 Meters

Sixty meters is the newest Amateur Radio HF band. It is also our only "channelized" HF band. Hams are only allowed to operate on five specific frequencies: 5330.5, 5346.5, 5366.5, 5371.5 and 5340.5 kHz. We're also limited to upper sideband voice (USB) communication at a maximum output of 50 W.

This band has much in common with 80 meters, but it is considerably quieter. A few other nations also share some of these frequencies, so DX contacts are possible—after sundown, of course.

40 Meters: 7 to 7.3 MHz

Forty meters is a transition band. It shares characteristics with the lower and higher HF bands.

During the day, 40 is excellent for communication over distances of about 500 miles or so. Phone operators enjoy meeting on 40 for late-morning chats. Nets—groups of hams who congregate on the air for a specific purpose—also exploit the advantages of 40 meters in the daylight hours.

At night, 40 meters opens to the world. CW operators enjoy global range in the lower portion of the band. SSB enthusiasts would relish the same contacts were it not for severe interference from shortwave broadcast stations. These high-powered RF blasters decimate the 40-meter voice segment throughout much of the world at night. The good news is that the ARRL has been working to persuade shortwave broadcast stations to move their 40-meter operations. You should see definite improvement by the year 2010.

30 Meters: 10.1 to 10.150 MHz

This is strictly a CW and digital band. No voice operating is allowed. In addition, you can't use more than 200 W output.

Thirty meters is good for DX work during the daylight hours, and up to several hours after dark. The CW operators gather in the lower portion of the band to chase international contacts. Thirty meters is also a terrific band for low-power (QRP) CW activity. The digital folks occupy the upper portion.

20 Meters: 14 to 14.350 MHz

Twenty meters is the queen of the DX bands. It's open to just about every corner of the world at various times of the day. During the peaks of the 11-year solar cycle, 20 meters will remain open throughout the night. Otherwise, it tends to shut down after local sunset. (The peak of the next solar cycle will likely occur in the year 2010.)

> **Twenty meters is the queen of the DX bands.**

All modes are hot and heavy on 20 meters! SSB reigns supreme in the upper half of the band. You may have a tough time finding a clear frequency between 14.225 and 14.350 MHz, especially on weekends. CW occupies much of the bottom portion. DX chasers often hunt between 14 and 14.050 MHz. Low-power (QRP) CW operators hang out around 14.060 MHz. The digital modes can be heard anywhere from 14.065 to 14.120 MHz.

17 Meters: 18.068 to 18.168 MHz

This daytime band also offers worldwide communication, though it is never crowded. SSB seems to be the dominant mode of communication on 17 meters, although you'll find a few CW stations and the occasional digital operator.

Seventeen meters is best during the peak years of the solar cycle. Even so, it can often provide global DX even in the worst years.

15 Meters: 21 to 21.450 MHz

This is a hot DX band when the solar cycle reaches its peak. In cycle doldrums, 15 meters can provide some mediocre DX, but it's usually limited to sporadic openings.

Like 17 meters, 15 is a daytime band. In the best years, it opens in the late morning and closes a few hours after dark. You'll find a mix of SSB and CW on this band, although the SSB operators usually predominate. The digital segment can also be quite active.

12 Meters: 24.890 to 24.990 MHz

When the solar cycle is up, so is this band—at least in the daylight hours. When the cycle is down, 12 meters is a wasteland. Even in the best years, contacts are few and far between. This has more to do with a lack of interest than anything else. You'll hear the odd SSB and CW conversation, but most operators prefer to migrate to the next-highest band—10 meters.

10 Meters: 28 to 29.700 MHz

During the best years of the solar cycle, 10 meters is one of the hottest DX bands around. At solar-cycle peaks, the ionosphere absorbs relatively little of your

> **During the best years of the solar cycle, 10 meters is one of the hottest DX bands around.**

signal at this frequency. It simply bends it back to Earth thousands of miles away. As a result, even low-power stations can use 10 meters to work the world with ease.

When sunspots are scarce, so are contacts on 10 meters. In fact, many hams consider 10 meters to be worthless during the solar minimum. That's an exaggeration. While it's true that you won't make too many DX contacts during the low points in the cycle, the band frequently opens for conversations over hundreds of miles.

When it's open, 10 meters is usually a daytime band. It opens in the late morning and shuts down at dark. SSB activity is most heavily concentrated in the segment from 28.300 to 28.500 MHz. CW is relatively rare and so are the digital modes. At the top end of the band you'll find a segment from 29.500 to 29.700 that's dedicated to FM operating.

THE WORLD ABOVE 50 MHZ

As you cross the invisible border at 50 MHz, you enter a new world where the Sun has much less influence. If you thought strange things happened on the HF bands, propagation is even stranger above 50 MHz.

6 Meters: 50 to 54 MHz

Some hams refer to 6 meters as the "magic band." The reason has much to do with its unpredictable nature. Most of the time 6 meters is limited to local and regional communication. When sporadic E comes into play, however, anything can happen. You can be listening to noise in your headphones one minute and hear voices from 1000 miles away the next minute! It takes very little power and only mediocre antennas to make DX contacts with sporadic E.

At the solar cycle peaks, 6 meters can also open for global propagation, usually at midday. In addition, contacts on paths between the US and South America are common.

Six meters is an excellent band for bouncing signals off the fiery trails of meteors (a type of propagation known as meteor scatter). And during solar storms when the aurora flares around the Earth's poles, 6-meter signals can reflect off these shimmering curtains and be heard at great distances.

2 Meters: 144 to 148 MHz

Like 6 meters, 2 meters is primarily a band for local and regional communication. Sporadic E can occasionally shake things up on this band as well.

Weather conditions begin to become important on 2 meters. The passing of a weather front can bring unusual "openings" over hundreds of miles, and sometimes farther.

As on 6 meters, hams enjoy chasing meteors and auroras on 2 meters. In fact, 2 meters is probably the most popular meteor-scatter band.

1.25 Meters: 222 to 225 MHz

This band has a lot in common with 2 meters. The difference is that sporadic-E propagation almost never makes it this high. Space communication is also difficult.

70 cm: 420 to 450 MHz

Now we've crossed into UHF. There is no sporadic E propagation whatsoever on this band. On 70 cm, weather rules. Most of the time, the band is quiet and useable for local contacts. When weather fronts move through, however, keep your ears open. Under these conditions, 70 cm can burst wide open for hundreds of miles. Even changes in air temperature are enough to trigger an "event" on this band.

33 cm: 902 to 928 MHz

There is little ham activity on this band, but it is capable of the same sudden, weather-related openings as 70 cm.

1200 MHz and Beyond

This is the microwave realm. In most cases you're looking at line-of-sight communication, but don't be fooled into thinking that DX is impossible. Again, weather has a major role to play on these bands. On the higher microwave bands, such as 10 GHz, it is even possible to bounce signals off raindrops!

WHEN ARE THE BANDS OPEN?

As you already know, the answer to this question depends on a number of variables. **Table 3-1** provides some very general recommendations for HF bands, but don't rely on them to set your schedule for you. Part of the fun of Amateur Radio is exploring the wonders of signal propagation and picking up valuable experience along the way.

The idea that a given band is "closed" can often be wrong—and contagious. If you have enough hams believing that, say, the 6-meter band is closed, no one will pick up their microphones and call CQ. So is the band really closed? It might as well be! Everyone is listening and no one is talking, which means that the granddaddy of all DX openings could come and go unnoticed. If you think the band is closed, call a few CQs just to be sure. A dead band can come alive when you least expect it!

Table 3-1

When are the Bands Open—General Guidelines

These are the times when you stand the best chance of making long-distance contacts. However, these are general rules only!

Band (MHz)	Open
1.8	Nighttime only
3.5	Nighttime only
5	Nighttime only
7	Nighttime only
10	Day and night
14	Day and night (around the clock at the solar cycle peak)
17	Daytime and early evening
21	Daytime and early evening
24	Midday through early evening
28	Midday through early evening
50	Midday through early evening

PAY ATTENTION TO THE SUN

If you're operating on frequencies below 50 MHz, it pay attention to what Old Sol is up to. A number of ham-oriented Web sites (such as DX Summit at **oh2aq.kolumbus. com/dxs/**) offer reports on the state of the Earth's geomagnetic field, which is directly influenced by the Sun. The more active the geomagnetic field becomes, the more HF propagation becomes unstable. When the activity becomes extreme, you can experience a total blackout of the HF bands. (You'll think your radio has failed!)

Geomagnetic reports include the so-called A and K indices. When the K index rises higher than 5, or when the A index exceeds 20, you can bet propagation is going to be uncertain. How bad can it get? When geomagnetic storms become severe, you may see the K index at 9 and the A index at 100. This means total blackout!

> *The idea that a given band is "closed" can often be wrong—and contagious.*

Think of the geomagnetic indices as barometers of HF propagation. The K index breaks down like this:

K0=Inactive

K1=Very quiet

K2=Quiet

K3=Unsettled

K4=Active

K5=Minor storm

K6=Major storm

K7=Severe storm

K8=Very severe storm

K9=Extremely severe storm

The A index is a broader measurement because it is an averaged number:

A0 - A7 = quiet

A8 - A15 = unsettled

A16 - A29 = active

A30 - A49 = minor storm

A50 - A99 = major storm

A100 - A400 = severe storm

Generally speaking, you want to see low A and K index numbers for good HF band conditions. A good day on the HF bands might sport an A index of 7 and a K index of 2.

Another number to pay attention to is the Solar Flux. This number is based on the amount of solar radiation measured on the 2.8 GHz band. It is closely related to the amount of ultraviolet radiation, which is needed to create an ionosphere. The higher the number, the better. At the bottom of the 11-year solar cycle, you may see Solar Flux numbers in the 80s or even 70s. At the cycle peak, the flux can exceed 200.

KEEP AN EYE ON THE WEATHER

If you're prowling the VHF, UHF and microwave bands for DX, keep your eyes on the sky (or on the weather forecasts). All of the Earth's weather takes place in the portion of our atmosphere known as the troposphere, and this is also where you'll find the DX action at 144 MHz and above.

If you're prowling the VHF, UHF and microwave bands for DX, keep your eyes on the sky (or on the weather forecasts).

Watch for temperature inversions after a weather front passes through your area. Be especially watchful for stable high-pressure zones ahead of cold fronts during warmer months. This is fertile ground for spectacular tropospheric band openings, often called *tropo* or *tropo ducting*. A good tropo band opening can carry VHF/UHF signals 1000 miles or more!

WEB TOOLS

Hams have put the Internet to good use as a means to share information about band openings. You'll find a variety of clusters (short for DX PacketClusters), dispensing news on an almost minute-by-minute basis. The way most of these clusters work is by

You'll find a variety of clusters (short for DX PacketClusters), dispensing news on an almost minute-by-minute basis.

posting announcements from hams who have found DX stations on the air. These reports are known as spots and look something like this:

N1RL   14082.7   J5UCW   RTTY

This spot translates to: "N1RL is hearing J5UCW on 14082.7 kHz using the RTTY digital mode."

There are several DX clusters. DX Summit at **oh2aq.kolumbus.com/dxs/** is one of the most popular. Many software-logging programs that you can purchase for your station computer include the ability to connect directly to these clusters. They'll even sound an alarm when a spot appears for a station you need for an operating award!

LISTENING FOR BEACONS

Hams take their DXing so seriously, they've established networks of propagation beacons throughout the world. These beacons are low-power transmitters that continuously send their call signs. Other beacons also send additional information such as their location, antenna type and so on. By listening for beacon signals, you can tell when a band is open, and where it is open to.

> *Beacons are particularly critical for monitoring propagation above 50 MHz since band openings can come and go so quickly.*

For HF work, the Northern California DX Foundation (NCDXF), in cooperation with the International Amateur Radio Union (IARU), maintains the most well-known HF beacon system. At the time of this writing, NCDXF/IARU beacons were on the air from 17 countries around the world. The beacons send their call signs in Morse code at 14.100, 18.110, 21.150, 24.930 and 28.200 MHz. See their Web page at **www.ncdxf.org/beacons.htm**l.

You'll also hear beacons on 10 meters, as well as on almost all the VHF, UHF and microwave bands. Beacons are particularly critical for monitoring propagation above 50 MHz since band openings can come and go so quickly.

No discussion of beacons would be complete without mentioning the powerful National Institute of Standards and Technology (NIST) radio stations.

Station WWV broadcasts time and frequency information 24 hours per day, 7 days per week to millions of listeners worldwide. WWV is located in Fort Collins, Colorado, about 100 kilometers north of Denver. The broadcast information includes time announcements, standard time intervals, standard frequencies, UT1 time corrections, a BCD time code, geophysical alerts, marine storm warnings and Global Positioning System (GPS) status reports. You'll hear WWV on 2.5, 5, 10, 15 and 20 MHz.

WWVH broadcasts the same information as WWV, but does it from the Island of Kauai, Hawaii.

The DX Summit at oh2aq.kolumbus.com/dxs/ is one of the most popular DX clusters in the world.

NIST station WWV in Colorado can be heard on several HF bands.

Listen for it on 2.5, 5, 10 and 15 MHz. Both WWV and WWVH are excellent tools for determining band conditions in a hurry.

There is yet another NIST station, but you won't hear it on a ham receiver. WWVB continuously broadcasts time and frequency signals at 60 kHz—that's way below the lowest ham band! Although you may never hear WWVB, you've probably seen products that depend on its signal. The next time you see an advertisement making a statement such as, "This timepiece never needs setting! It synchronizes itself to an atomic clock!", they're talking about the fact that the device includes a tiny radio receiver tuned to WWVB on 60 kHz. You can learn more about the NIST stations on the Web at **tf.nist.gov/timefreq/index.html.**

4 Using Your Voice on the HF Bands

After you've assembled and tested your station, there is nothing left to do but get on the air. The best approach by far is to spend several hours *listening* before you reach for the microphone. Tune through the bands and eavesdrop on as many conversations as possible. When you finally feel comfortable with the territory, it's time to throw some RF!

Whenever you transmit, you're representing all of Amateur Radio.

Your transceiver manual is the best source of information on how to properly set up your rig for voice operating. Still, some general rules apply…

■ Learn to use your microphone. Start by positioning the microphone about one inch from your lips. Speak in a normal tone of voice.

■ Most transceivers have an *ALC* (Automatic Limiting Control) meter. Select this meter and watch it as you speak. If the meter indicates that your voice is bouncing it out of the ALC range, you'll need to find the microphone gain control and turn it down. Alternatively, try speaking a little softer. A high ALC reading indicates that you are overdriving your radio and possibly distorting your signal.

Randy, K5ZD, enjoys contest operating.

■ Hams usually switch their radios from transmit to receive (and back again) by pressing a button on the microphone (known as Push To Talk, or *PTT*), or by using a foot switch. For hand or desk microphones, avoid stabbing or punching the switch. Not only does this shorten the life of the switch, it can send a loud *click* at the beginning of your transmission.

■ Nearly every modern radio has a *VOX* function—Voice-Operated Switch. When the VOX is on, the sound of your voice will automatically switch the transceiver into the transmit mode. This sounds like a fine thing, and when used properly, it is. However, it is easy to abuse VOX, with annoying consequences. If you must use VOX, speak into the microphone at a normal voice level. When you start speaking, the transmitter should activate automatically. When you finish speaking, the transceiver should return to the receive mode (after a short delay). Sometimes the VOX may trigger in response to background sounds. If so, look for the VOX Gain control. You can adjust this control to eliminate the problem.

A speech processor takes a normal voice signal, which varies constantly as you speak, and processes the signal to minimize fluctuating power levels.

■ You'll also find *speech processing* (or *speech compression*) as a feature on most radios. In a nutshell, this is a method of boosting your average output power when you're operating SSB. A speech processor takes a normal voice signal, which varies constantly as you speak, and processes the signal to minimize fluctuating power levels. The result is an SSB signal that has consistent power at the highest level possible.

Speech processing can work to your advantage when your signal is weak, such as when you are operating at low power, or with a poor antenna. On the other hand, there is no such thing as a free lunch. Speech processing can distort your signal, sometimes severely. If you use speech processing, keep the processing level set at medium and ask for reports on your signal quality. Turn the level down if other operators tell you that your signal is distorted.

GETTING STARTED

To get a voice chat off the ground, you have two choices: You can call CQ, or you can answer someone who is calling CQ.

Before calling CQ, it's important to find a frequency that appears unoccupied by any other station. This may not be easy, particularly in crowded band conditions. No matter what mode you're operating, *always listen before transmitting*. Make sure the frequency isn't being used *before* you come barging in. If, after a reasonable time, the frequency seems clear, ask if the frequency is in use, followed by your call.

> No matter what mode you're operating, always listen before transmitting. Make sure the frequency isn't being used before *you come barging in.*

"Is the frequency in use? This is NY2EC." If nobody replies, you're clear to call.

Keep your CQ very short. Longwinded CQs drive most hams crazy. Besides, if no one answers, you can always call again. If you call CQ three or four times and don't get a response, try another frequency.

A typical SSB CQ goes like this:

"CQ CQ Calling CQ. This is AD5UAP, Alfa-Delta-Five-Uniform-Alfa-Papa, calling CQ and standing by." Notice how the call uses standard *phonetics* (see **Table 4-1**). We often use phonetics to make it easier for other operators to decipher our call signs. Phonetics are especially helpful when the other station is not hearing you well, or when the operator is not a fluent speaker of English.

With this in mind, *don't* make up your own phonetics. For example, a ham might say "Alfa-Delta-Five-Up-All-Night." This is clever and amusing, but it doesn't help the person who is trying to figure out your call sign.

Table 4-1					
Standard Phonetic Alphabet					
A	Alfa	J	Juliett	S	Sierra
B	Bravo	K	Kilo	T	Tango
C	Charlie	L	Lima	U	Uniform
D	Delta	M	Mike	V	Victor
E	Echo	N	November	W	Whiskey
F	Foxtrot	O	Oscar	X	X-ray
G	Golf	P	Papa	Y	Yankee
H	Hotel	Q	Quebec	Z	Zulu
I	India	R	Romeo		

And if you're answering a CQ, keep the answer short as well. Say the call sign of the station once or twice only, followed by your call repeated twice.

"N2EEC N2EEC, this is AB2GD, Alfa-Bravo-Two-Golf-Delta."

CHEWING THE RAG

"Rag chewing" is ham lingo for a long, enjoyable conversation. Just start with the basic facts: your name, location, his signal report, and possible a brief summary of your station (how much power you're running and the kind of antenna you're using).

Once you're beyond the preamble, the topic choice is yours. The tried and true formula for success is to get the other person to talk about himself. Any life worth living has at least *one* interesting aspect. You may have to dig this aspect out of your palaver partner, but it's often worth the effort. If all else fails, make the following request:

"Look out the window and tell me, in detail, exactly what you see."

You'll definitely throw the other person off guard—that much is guaranteed! If they're in a room without a window, don't let them off the hook. "What would you see if you *did* have a window?"

Hams can talk about anything, but there are some topics we try to avoid. Discussions of politics and religion tend to attract controversy and start arguments on the air. If it looks like your rag chew is heading in those directions, use good judgment. Does the other operator agree with your views? If not, will you be offended? Will he (or others) be offended? If you have doubts, it is best to change the subject. Conduct yourself as though anyone in the world might be listening at any time. Whenever you transmit, you're representing all of Amateur Radio.

SHORT AND SWEET

Hams love to talk, but there are times when you should keep the conversation short.

For instance, you may run into a DX operator who doesn't have a good command of English. He may be reading from a "script" of standard English sentences and may not be able to carry on a complete conversation. Listen before you call. If he seems to be making short contacts (signal report and

If conditions are favorable, you may end up having a conversation with JP1NWZ in Japan.

"good-bye"), it is best to do the same.

This is especially true if the DX station is working a *pileup*. This is something that we'll talk more about in another chapter. Suffice to say that this is a situation where the DX station is trying to contact as many people as possible, as rapidly as possible.

You'll hear a torrent of signals as other hams try to make quick contacts. If you are lucky enough to be heard by the DX station, only give the information he is looking for—usually a signal report and your location. *Do not* attempt to engage him in conversation. This will make you very unpopular with the DX operator, as well as your fellow hams!

You may occasionally hear hams operating *special event stations*. These are temporary stations set up at events throughout the country. Hams establish special event stations at county fairs, boat races, or just about any event or occasion you can think of. (You can do this too, by the way!) Special event stations are fun to contact and many offer special certificates or contact confirmation cards (known as *QSLs*). But unless the special event operator sounds like he or she wants to indulge in chit-chat, keep the conversation short. This gives everyone else a chance to make their contacts.

If you happen to hear stations making calls like this…

"CQ Contest! CQ Contest! This is W1AW. Contest!"

… you've stumbled across an on-the-air contest. We'll discuss contests in another chapter, but the idea is simple: contact as many stations as possible during the contest period. You don't have to be involved in the contest to participate; your contact will count regardless. Just listen to the contest station and determine what he is looking for. Contest contacts require the exchange of specific information such as your state, county, etc. Find out what he needs *before* you call and make the contact short. Time is essential when it comes to a contest!

JOINING A CONVERSATION IN PROGRESS

If you don't want to call CQ to start a conversation, the alternative is to join a conversation that is already in progress. Remember that private conversations don't exist in Amateur Radio. Anyone can join a two-way chat and make it a three-way.

The key to joining a conversation is to use proper technique. Listen carefully to the operators. Are they having an animated, involved discussion? If so, it may be a bit rude to interrupt unless you have something important to offer. For example, if they are talking about a problem they are

FM on the HF Bands?

Yes indeed, there are frequency modulated voice transmissions—*FM*—on the HF bands, but only on 10 meters above 29.500 MHz. When the 10-meter band is popping at the peak of the 11-year solar cycle, you'll hear a great deal of 10-meter FM activity. Listen for direct station-to-station conversations (*simplex*) at 29.600 MHz. There are also FM repeaters on 10 meters. These are automatic stations that relay signals. You transmit on the repeater *input frequency* and listen on the *output frequency* (we'll talk about FM and repeaters in another chapter). Here are the common input/output repeater frequency pairs for 10-meter FM…

Input (MHz)	Output (MHz)
29.520	29.620
29.530	29.630
29.540	29.640
29.550	29.650
29.560	29.660
29.570	29.670
29.580	29.680
29.590	29.690

Digital Voice

What is that raspy buzzing sound? If you're hearing it in the voice portion of the band, it might be a digital voice transmission.

Hams have been using digital voice transmissions for several years. They are not common on the HF bands, but you will hear them from time to time. The advantage of digital voice is the clear audio quality, along with the ability to send other data (such as text or even small images) while you are talking. The disadvantage is that digital voice usually requires strong signals to work well. If the signal becomes too weak, the audio signal will abruptly stop. Unlike an AM, FM or SSB signal, a digital signal is either audible 100% or it isn't audible at all.

When this book went to press, there were two ways to create an HF digital voice signal. One way was to purchase a digital voice modem such as the AOR AR9800. This device connects between your microphone and your radio, turning your voice into rapidly changing audio tones that can be transmitted over the air. The digital modem also decodes the signals that you receive.

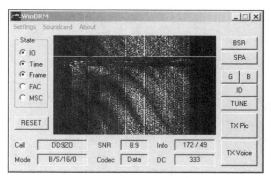

WinDRM software uses your computer sound card to send and receive digital voice transmissions.

The other method is to use a computer sound card and a piece of *Windows* software known as *WinDRM*. The software acts in the same fashion as the hardware modem, changing your voice into tones for transmission and decoding the received signal back into analog audio. To use *WinDRM* you'll need a sound card interface such as those manufac-

The AOR ARD9800 digital voice modem (the small box with the microphone connected to it).

tured by West Mountain Radio, MFJ Enterprises, MicroHAM, MixW RigExpert or TigerTronics. See the advertising pages of *QST* magazine. The *WinDRM* software is free and you can download it from **n1su.com/windrm/**.

Digital conversations typically begin in SSB. The operators establish contact, and then switch to digital. If the digital link fails, they return to SSB.

having with a particular radio and you can offer helpful advice, by all means join in!

If the chat seems casual and rambling, the operators may not mind someone else joining the roundtable. Sometimes a good conversationalist—such as yourself!—is quite welcome.

So how do you politely interrupt? The best approach is to wait for one station to stop transmitting and then quickly announce your call sign.

"Yes, Charlie, I hope to get the dipole soldered together this weekend if the weather holds up."

"N1RL."

"Ah…we have a breaking station. N1RL, go ahead!"

Alan, K6SRZ, was part of a ham expedition (known as a *DXpedition*) to Kure, a tiny Pacific island. Alan's goal as a highly desired DX station was to make as many contacts as possible.

Do not use the word "break." You'll hear other hams doing this, but it is bad operating practice. You should only say "break" when you need to interrupt because of an emergency.

JUMPING INTO THE NETS

Whenever you have a group of hams who meet on a particular frequency at a particular time, you have a net. There are nets devoted to just about every purpose you can imagine. If you're a user of Kenwood equipment, there is a Kenwood user's net. If you're into amateur satellites, you can hang out with the AMSAT nets. DX nets are devoted to arranging contacts with difficult-to-work stations. *Traffic handling* is a popular net activity. Hams meet on various frequencies to relay messages from throughout the nation or the world. Other nets are simply groups of hams who like to meet and talk about whatever is on their minds.

A net would dissolve into chaos if it wasn't for the *net control station*, or NCS. He or she acts like a traffic cop at a busy intersection. The NCS "calls" the net at the appropriate time. If you happen to tune in at the start of an SSB net, you might hear something like this . . .

"Attention all amateurs. Attention all amateurs. This is KX6Y, net-control for the Klingon Language Net. This net meets every Saturday at 1400 UTC on this frequency to exchange news and views concerning the language of the Klingon homeworld. Any stations wishing to join the net please call KX6Y now."

That's the cue for any interested ham to throw out his or her call sign. The NCS writes down each call sign in the order in which he receives it. At some point he may break in and say, "Okay, so far I have WB8IMY, N6ATQ, WR1B, KB1EIB and K1ZZ. Anyone else?"

Once the list is complete, the NCS will call each station in turn and ask if they have any questions or comments for the net.

Once the list is complete, the NCS will call each station in turn and ask if they have any questions or comments for the net. If he calls you, you have the option of speaking your piece or telling the NCS to skip to the next person. The NCS will also ask for new check-ins from time to time. If you didn't join the net at the beginning, that's your chance.

Traffic and other public-service nets are more tightly organized and follow stricter rules concerning who can say what . . . and when. For more information about these types of nets and how they work, get on the Web and go to **www.arrl.org/FandES/field/pscm/**. You'll also find a directory of Amateur Radio nets at **www.arrl.org/FandES/field/nets/**.

5 Code Conversations

A classic "straight" key.

How's this for an overbearing description?

"Continuous wave radiotelegraphy using International Morse code."

Sheesh! No wonder hams just call it "CW."

CW is the oldest form of radio communication. It consists of nothing more than turning a signal on and off at regular intervals according to an established code system. For hams, that code system is known as International Morse code—the "Morse" being Samuel F.B. Morse, the man who invented the code more than 150 years ago.

Sam was one of those troublesome geniuses who kept beating people over the head with his idea until someone paid attention. His crazy scheme was to send information at the speed of light over a wire (wireless was still 50 years away), and he figured he could do it by merely switching an electric current on and off. Sam created a code consisting of short current pulses (dits) and longer pulses (dahs). With this arrangement, he coded the entire English alphabet, the numbers 0 through 9 and a bit of punctuation.

Morse's system was a natural fit for radio because of its simplicity. The pioneers of radio were using extremely crude equipment and they needed a simple method to communicate information. What could be easier than just turning the signal on and off?

CW TODAY

There was a time when CW was the undisputed king of Amateur Radio. But as voice modes such as AM, FM and SSB grew in popularity after World War II, the use of CW declined. It declined further with the advent of digital ham communication.

Even so, CW remains stubbornly popular for several reasons:

> *CW is the only mode that can be used for communication under poor signal conditions without a computer.*

■ CW signals can be heard under poor conditions when voice communication is impossible and even digital signals have difficulty.

■ CW transceivers are much simpler to design and build than voice or digital radios. Very few hams still build voice transceivers, but many continue to build their own CW radios.

■ Unlike the digital modes, CW does not require a computer to decode. In fact, CW is the *only* mode that can be used for communication under poor signal conditions without a computer.

■ A CW signal is highly spectrum efficient, which means it only occupies a very small bandwidth.

CW is also a uniquely human skill, not unlike mastering a musical instrument or learning a foreign language. It takes some effort to learn CW and still more effort to become proficient. As any CW enthusiast will tell you, however, the quest is well worth the reward. Mastery of CW opens a whole world of enjoyment that is closed to everyone else.

LEARNING CW

It takes practice to learn CW. Try to set aside 20 minutes each day to do nothing more than listen. The ARRL sells audio CDs with Morse code practice at various speeds. You'll find these for sale on the ARRL Web at **www.arrl.org/catalog**, or you can call the League toll free at 1-888-277-5289.

The ARRL Headquarters station, W1AW, also transmits Morse code practice on weekday evenings. If you have a radio, this is a great way to copy code under real signal conditions. You'll find the W1AW schedule on the Web at **www.arrl.org/w1aw.html**.

Regardless of which method you choose, start slow. Try copying at 5 WPM (words per minute), writing down the letters, numbers and punctuation (see **Table 5-1**) as you hear them. Don't think too long about each character—make your best guess as quickly as possible. At first, most of your guesses will be wrong, but as you continue practicing you will notice that the number of correct guesses will increase.

Practice at one speed for a week or two. As you gain confidence, challenge yourself by trying faster speeds. Also, start listening to live CW conversations on the air. This will give you a feel for how they are conducted. A couple of weeks into your practice routine try to just listen and decipher the code *without writing it down*. This will be difficult— *guaranteed*—but it is one of the important steps on the road to absolute Morse code fluency. Most of the hams who've mastered Morse at higher speeds never write down what they hear, letter for letter. They listen to the code in the same way that you'd listen to speech.

Table 5.1

International Morse Code

· = *dit*   – = *dah*

Alphabet

·–	A	––·	G	––	M	···	S	–·–– Y
–···	B	····	H	–·	N	–	T	––·· Z
–·–·	C	··	I	–––	O	··–	U	
–··	D	·–––	J	·––·	P	···–	V	
·	E	–·–	K	––·–	Q	·––	W	
··–·	F	·–··	L	·–·	R	–··–	X	

Numbers

·––––	1	–····	6
··–––	2	––···	7
···––	3	–––··	8
····–	4	––––·	9
·····	5	–––––	0

Punctuation marks

Point (.)	·–·–·–
Comma (,)	––··––
Question-mark (?)	··––··
Colon (:)	–––···
Hyphen (-)	–····–
At-sign (@)	·––·–·
Error	········
Slant (/)	–··–·

SETTING UP YOUR CW STATION

Every ham transceiver for the HF bands includes CW among its operating modes. So do a number of so-called multimode transceivers for VHF and UHF. All you need to send CW is a device to key your radio on and off.

The most basic is the *straight*

Practice at one speed for a week or two. As you gain confidence, challenge yourself by trying faster speeds.

This is how most paddle key operators position their fingers. Depending on how the paddles are set up, a light touch of the thumb sends a dah while the index finger sends a dit.

key, and it is as old as Morse code itself. Learning to use a straight key well isn't easy; it takes practice to send smooth CW. Most hams can send up to 15 WPM with a straight key, although some can send at much higher speeds.

The most popular method of sending CW is with the *electronic keyer*. This usually consists of a set of finely balanced switches known as *paddles*. Hams who've mastered electronic keyers can send CW at astonishingly high speeds. It is amazing to watch their fingers dance on the paddles!

The paddles are connected to the electronic keyer, which automatically generates dits and dahs depending on which paddle you touch. If you touch the dah paddle, for example, the keyer will send a string of dahs until you release the pressure; touch the dit paddle and dits are sent. To send the letter "Q," you'd touch the dah paddle just long enough for the keyer to send two dahs, touch the dit paddle just once, and then finally touch the dah paddle again to send the last dah: *dah-dah-dit-dah*. The keyer has an adjustment so that you can select the sending speed you desire.

Many modern transceivers have the electronic keyer built in. If your radio doesn't include this feature, you can purchase an external keyer.

It is also possible to send CW with a computer keyboard. This is anathema to CW purists, though, and is usually only done by hams involved in contests where the same information is sent repeatedly.

> *The most popular method of sending CW is with the electronic keyer.*

STARTING A CONVERSATION

The best way to start a CW chat is to tune around until you hear someone calling CQ. CQ means, "I wish to contact any amateur station." In time you'll learn to recognize the sound of a CQ call. It has an unmistakable rhythm!

When answering a CQer you should *zero beat* the other ham's frequency. That means setting your transmit frequency as close to theirs as possible. Not only does this technique conserve space by keeping signals close together, it can also have a big influence on whether the CQer can hear your call. The reason for this is that some stations use narrow receive filters. If the CQer has his narrow filter activated, he may not hear you answer if you are more than a few hundred cycles away from his transmit frequency.

So how do you zero beat? One method is to tune down through the other CW signal, the pitch going from high to low, until the other signal disappears. Then,

Q-Signals

CW operators use Q-signals to quickly communicate certain ideas. Q-signals are particularly well suited for use when you're talking with someone who doesn't speak English. Although Q-signals should only be used for CW, you'll often hear voice operators using Q-signals as well.

QRL	Is the frequency busy?	QRT	stop sending
QRM	interference	QRX	wait, standby
QRN	noise, static	QSB	fading
QRO	increase power	QSL	acknowledge receipt
QRP	decrease power	QSY	change frequency
QRS	send slower	QTH	location

CW Abbreviations

CW operators often use abbreviations to make communication more efficient. Here are some of the more common abbreviations you are likely to hear on the air.

ADR	address	GN	good night	RIG	station equipment
AGN	again	GND	ground	RPT	repeat
BK	break	GUD	good	SK	end of transmission
BN	been	HI	laughter	SRI	sorry
C	yes	HR	here	SSB	single side band
CL	closing	HV	have	TMW	tomorrow
CUL	see you later	HW	how	TNX-TKS	thanks
DE	from	N	no	TU	thank you
DX	distance	NR	number	UR	your
ES	and	NW	now	VY	very
FB	fine business	OM	old man	WX	weather
GA	go ahead	PSE	please	XYL	wife
GB	good bye	PWR	power	YL	young lady
GE	good evening	R	roger	73	best regards
GM	good morning	RCVR	receiver	88	love and kisses

slowly tune back up until you hear a pleasant CW note in your receiver (typically 600 Hz). This should put your signal close enough.

An alternative is to use the *Receiver Incremental Tuning* (RIT) feature found on many transceivers. (On some radios this may be called a *clarifier*.) With the RIT *off*, tune down into the signal until it disappears. Then, turn the RIT on and adjust until you hear the signal again. The RIT adjusts your receive frequency slightly, but leaves your transmit frequency unchanged.

> **When answering a CQer you should zero beat the other ham's frequency.**

If you can't find anyone calling CQ, perhaps you should try it yourself. A typical CQ goes like this: CQ CQ CQ DE KD4AEK KD4AEK KD4AEK K. The letter K is an invitation for any station to reply. If there is no answer, pause for 10 or 20 seconds and repeat the call. If your transceiver has a narrow CW receive filter, it is a good idea to turn the filter off when calling. Don't send faster than you can receive. By the same token, don't respond to a CQ at a speed faster than the other station is sending.

If you hear a CQ, wait until the ham finishes transmitting (by ending with the letter K), then call him. Make your call short, like this: K5RC K5RC DE K3YL K3YL AR (AR means "end of message").

Suppose K5RC heard someone calling him, but didn't quite catch the call because of interference (QRM) or static (QRN). Then he might come back with QRZ? DE K5RC K (Who is calling me?).

Prosigns

A prosign is a special character that consists of two letters sent together with no pauses in between. Here are a few common ones…

Wait, stand by (AS)	dit dah dit dit dit
Slash (DN)	dah dit dit dah dit
End of message (AR)	dit dah dit dah dit
End of contact (SK)	dit dit dit dah dit dah
Break (BT)	dah dit dit dit dah
Back to you *only* (KN)	dah dit dah dah dit
Closing Station (CL)	dah dit dah dit dit dah dit dit

THE CONVERSATION IS UNDERWAY

Most HF contacts begin with an exchange of basic information: Names, locations, equipment, signal reports and even weather reports. After that, it's up to you. Sometimes you'll find that

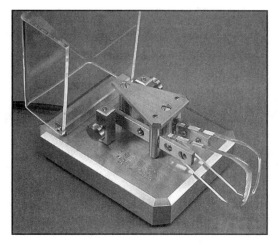

This is a precision paddle set.

If your radio doesn't have an electronic keyer built in, you can add one externally, such as this K1EL WinKeyer.

Bernie, W3UR, enjoys making DX contacts using CW.

The most common method to correct a mistake is for the sending station to send a rapid series of dits, followed by the character or word you meant to send.

you have to draw the other person into the conversation. The best way to do that is to ask questions. For example, ask what the person does for a living. She's a doctor? Okay, ask about her specialty, where she practices and more. In other words, get her to talk about herself. If you ask the right questions, the conversations will unfold on their own.

During the contact, when you want the other station to take a turn, the recommended signal is $\overline{KN}$, meaning that you want only the contacted station to come back to you. If you don't mind someone else signing in, just K ("go") is sufficient. You don't need to identify yourself and the other station at the beginning and end of every transmission. That wastes time. The FCC only requires you to identify *yourself* every 10 minutes.

Everyone occasionally makes a mistake while sending. The most common method to correct a mistake is for the sending station to send a rapid series of dits, followed by the character or word you meant to send. You will hear some hams use a question mark to signify that they are going to repeat a word, even if they haven't made a mistake. For example, "MY NAME IS STEVE? STEVE." This use of a question mark is frequently employed to indicate the repetition of a difficult or unusual word.

ENDING THE CONVERSATION

When you decide to end the contact, or when the other ham expresses his/her desire to end it, don't keep talking. Briefly express your thanks: TNX QSO or TNX CHAT— and then sign out: 73 $\overline{SK}$ WA1WTB DE K5KG. If you are leaving the air, add $\overline{CL}$ to the end, right after your call sign.

6 The Digital Universe

The heart of any Amateur Radio digital communications station is the computer sound card.

Hams don't just use their voices or code keys to communicate. We've been swapping digital information for decades, long before it was fashionable or popular to do so. In fact, Amateur Radio digital communication is much older than the Internet.

Hams use digital communication for casual conversations similar to Internet instant messaging or chat rooms. Most ham chats involve the exchange of text, but some modes allow images to the swapped as well. As you'll learn in the next chapter, there are also on-the-air digital contests and even challenging digital DX pileup competitions.

Nearly all Amateur Radio digital communication is built around a single piece of hardware: the computer sound card. The great thing about sound cards is that nearly every modern PC has one—either as an actual stand-alone card or a built-in sound chipset on the motherboard. Combine these sound systems with software and a simple computer-to-transceiver interface and you're on the air. You can hop from one mode to another by simply changing software. It doesn't get much easier than that!

> *Nearly all Amateur Radio digital communication is built around a single piece of hardware: the computer sound card.*

Let's take a brief look at the top-five most popular amateur digital communication modes, in no particular order.

RTTY (radioteletype, pronounced "ritty") is the granddaddy of digital hamming. Born just after the end of World War II, RTTY was the only amateur digital mode for more than 30 years. Although RTTY has lost its exclusive status, it is still the mode of choice for digital contesting and DXing.

Packet radio has been in existence since the early '70s and hams embraced it with gusto in the middle '80s. Packet is an error-detecting mode, which means that it is capable of communicating error-free information, including binary data (for images, software applications, etc). You'll find a bit of packet radio on the HF bands, but most packet activity takes place on VHF and UHF.

PACTOR is one of the only digital modes that requires specialized hardware rather than a computer sound card. PACTOR strolled onto the telecommunications stage in 1991.

It combined the best aspects of packet (the ability to pass binary data, for example) and robust error-free technology. PACTOR comes in three versions: I, II and III. PACTOR is the number one mode for mailbox operations and other forms of message handling on the HF bands.

PSK31 could be viewed as a high-octane cousin of RTTY. It is not an error-free digital mode, but it offers excellent weak-signal performance. It is the most popular mode for casual conversation. With PSK31 you can do wonders with low power (as low as 5 watts) and a minimal antenna. There are PSK31 enthusiasts who have contacted stations throughout the world with 20 watts and wire antennas in their attics!

SSTV, or slow-scan television, is not necessarily a digital mode. In its traditional form an SSTV signal is comprised of shifting audio tones, but the tones don't constitute a specific digital code. However, newer forms of this mode are truly digital.

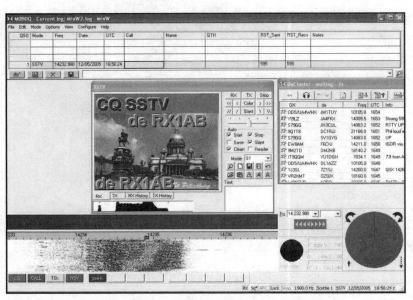

An SSTV image received with *MixW* sound card software.

None of the ham "TV" modes on the HF bands can send moving pictures; they only carry still images. Hams *can* send full-motion TV images, but this activity must take place on UHF and microwave frequencies because of the bandwidth required.

Remember: These are only the *most popular* amateur digital modes. There are many more with names like Hellschreiber, Olivia, Domino, MFSK16 and MT63. If you'd like to learn more, pick up a copy of the *ARRL HF Digital Handbook*.

BUILDING YOUR DIGITAL STATION

In the old days of digital, back when RTTY was your only option, setting up a station wasn't a trivial exercise. You had to make room for a bulky mechanical teletype machine, cobble together an interface to your radio to actuate the teletype, and install an oscilloscope to help you tune the signal. You sent text by typing on the teletype's awkward "green keys," and read the other station's replies on sheets of yellow paper.

Thanks to personal computers and microprocessor technology it is infinitely easier to assemble a digital station today. The only hurdle is to bridge the gap between the digital world of your computer and the analog world of your transceiver.

> *Thanks to personal computers and microprocessor technology it is infinitely easier to assemble a digital station today.*

Digital to Analog—And Back Again

Data exists in your computer in the form of changing voltages. Five volts might represent a binary "1" while zero volts may represent a binary "0." But a radio can't transmit changing voltages—at least not without a little translation help.

A collection of RigBlaster sound card interfaces by West Mountain Radio.

In the case of digital communication, our translator is a modulator/demodulator, otherwise known as a *modem*. A modem takes the binary data from your computer and translates it into modulated audio that can be sent over a radio. This audio consists of tones that change their frequencies, phase, amplitude and so on.

RTTY, for example, communicates 1s and 0s by creating a signal that shifts rapidly between two frequencies. When the shifting signal is an audio tone applied to a transceiver, the technique is known as audio frequency shift keying, or *AFSK*. You'll often see transceiver manuals referring to this as frequency shift keying, or *FSK*, and this can be a little confusing. True FSK involves changing the frequency of your rig's master oscillator from one *RF* frequency to another in sync with the raw data from your computer (the data isn't translated into an audio frequency signal).

PSK31, on the other hand, communicates binary data by shifting the phase of a single audio frequency, which is then applied to the transceiver. This is known as *phase shift keying*. The phase shifts correspond to binary 1s and 0s.

But what about reception? The modem waiting patiently at the other end of the path is equipped with signal decoders and sharp audio filters—electronic or digital. It will only respond to signals *if* they have the proper characteristics. Off-frequency signals are ignored, and signals separated by incorrect shifts or using incorrect modulation never make it past the filters. But when the tones hit the targets, the modem instantly converts them into data pulses—which soon wind up in your computer. Your computer software takes it from there.

Soundcards as Digital Modems

Sound cards do exactly what modems do: Convert data to audio and audio to data.

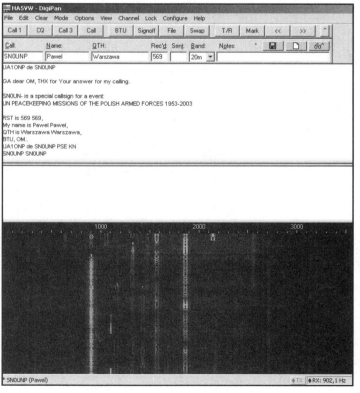

This is an actual PSK31 conversation in progress. The waterfall display is in the lower third of the image. To tune a signal, just click your mouse on any line in the waterfall.

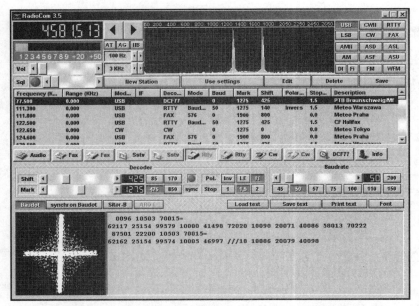

This is a RTTY signal being received and decoded with sound card software. The two sharp spikes in the top portion of the display are the mark and space tones. They are also represented by the crosshair-tuning indicator in the lower left corner.

With the proper software running in the background, your PC can become a high-performance digital communication machine.

The concept of connecting a sound card to your radio is simple. Sound comes out of your radio and into your computer to be processed, or decoded, in the receive path. Processed sound comes out of your computer and goes into your radio to be modulated and transmitted. The input to the sound card is the receive (audio) signal and the output is the transmitted (audio) signal. Most of the work is done (inside the computer) with the incredible software available to us.

Sound cards come in lots of flavors: 8 bit, 16 bit, 32 bit, wave table synthesis, ISA, PCI, gold, 128, "surround sound" and so on.

As we mentioned previously, a computer can, instead, use sound processing chips on the motherboard. That is usually the case with laptops. There are some new external sound adapters that function as a sound card, but are connected externally through a USB port. If your computer has sound it will have at least a headphone, microphone or speaker jack. If so, you can use it with your radio to communicate. A complex sound card is not needed—a modern, though inexpensive PCI sound card will work very well and a simple one will be easier to connect and operate. Older, ISA bus cards will work with most Amateur Radio programs but might have limited dynamic range and slower processing.

Even the simplest of sound cards can be complicated. They can have as few as two external connections but there may be as many as twelve or more. At the rear of your computer you may find LINE IN, MIC IN, LINE OUT, SPEAKER OUT, PCM OUT, PCM IN, JOYSTICK, FIREWIRE, S/PDIF, REAR CHANNELS or SURROUND jacks, just to name a few. Connections appear not only on the outside, but inside, as well. The main internal connection is to the ISA or PCI bus and this is by edge connection to the board. There may also be CD audio, telephone, daughterboard and PC speaker connections inside and on the board, as well.

The first thing to do is to identify which connection is which, at least at the card edge bracket, outside the board. The jacks are usually poorly marked with nonstandard international symbols or color codes.

Depending on your computer, you may be able to choose your receive audio connection from either MIC INPUT or LINE INPUT. Anything else you find is not an audio input. If you do have a choice, use the LINE INPUT for the receive audio from your radio. Although the MIC INPUT jack can be used, it will have much more gain than you need and you may find adjustment quite critical. The MIC INPUT jack will work better, however, if your sound card has an "advanced" option to select a 20 dB attenuator. You should not need ground isolation for receive audio as it is at a high level and is normally not susceptible to ground loops. You may also be able to choose from several outputs that appear on the radio. Your radio may have SPEAKER, HEADPHONE, LINE OUT, RECORD,

PHONE PATCH and DATA OUTPUT jacks available. These may be fixed output or variable output (using the radio's volume control). Be careful with radio's DATA OUTPUT jack—it may not work on all modes.

For the transmit connection, you will have a choice of the computer's HEADPHONE OUTPUT, LINE OUTPUT, SPEAKER OUTPUT or a combination of these. The SPEAKER OUTPUT is usually the best choice as it will drive almost anything you hook to it. The SPEAKER OUTPUT has a very low source impedance making it less susceptible to load current and RF. Any one of these outputs will usually work fine, provided you do not load down a line or headphone output by using very low impedance speakers or headphones to monitor computer-transmitted audio. The transmit audio connection must have full ground isolation through your interface, especially if it drives the MIC INPUT of the radio.

Again, DATA INPUT jacks may not work the way you expect. Never use the digital input or output of your sound card. Even though you may wish to operate a digital mode, it will not work. You must use an analog audio connection, not a digital one. These jacks may be labeled PCM AUDIO and they have a digital data stream coming out of them, but it's not audio. Currently manufactured radios require analog audio.

Depending on your computer, you may be able to choose your receive audio connection from either MIC INPUT or LINE INPUT. Anything else you find is not an audio input.

Setting Levels

Setting the sound card levels correctly is an area in which you may run into difficulty. Sound cards are much more complex than you might think; most sound cards have more than 35 separate adjustments supported by two major virtual control panels. No two computers are alike, and there are always several different ways to get the same results. To make matters worse, once you adjust your sound card don't expect it to stay adjusted. Running other programs and rebooting your computer may reset your sound card parameters.

You may find that you have to readjust your card the next time you need to use it for communications. Sound card control is accessed through several virtual control panels. The sound card may also be controlled automatically through the software programs that you use. There can be many analog and digital signal paths in a typical sound card. Your job is to identify which ones you are connected to and make those connections work. When you single-click (left) the speaker icon in the bottom right corner of your computer screen you get one simple up/down slider with a mute switch. If you double-click that same icon, however, you get a large control panel for playback as shown in **Figure 6.1**.

Your control panel may be different. Instead of saying "Volume Control" at the top left corner, it may say "Playback" or it may have

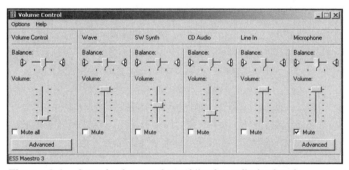

Figure 6.1—A typical sound card "volume" playback control panel.

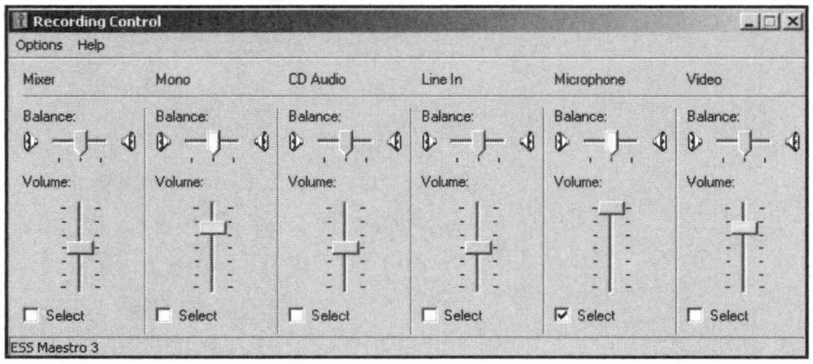

Figure 6.2—A sound card "record" audio panel.

"Select" instead of "Mute" boxes at the bottom. Each section has a label indicating the control for a single mono or stereo signal path: Volume Control, Wave, MIDI, CD Player, etc. The word Balance that appears to be part of a section label is not; it is just a label for the left/right slider directly beneath it. This is the playback, output or monitor control panel or, as hams would call it, the transmit control panel. This panel's only function is to adjust what comes out of your sound card. The only exception would be if you had an older sound card—the controls might be ganged with the recording control panels; they then could not be adjusted independently. **Figure 6.2** shows a typical "Recording" control panel. This panel is carefully hidden in *Windows* and you may not have known it was there unless you are already operating the sound card modes. Here's how to find it: double click the speaker icon, then "Options," "Properties," "Recording" and finally, "OK." Surprise…another control panel appears! The "Recording Control" panel is used to turn on and adjust the input signals from both analog (audio) and digital inputs and to feed those input signals to the software for processing. This functions as a receive control panel. It does not normally control transmit unless you are transmitting an audio signal that is being fed from your radio, through your computer and software, and processed in real time.

An example would be microphone equalization or speech processing. Depending on your sound card there may be buttons for "Advanced" settings. Here you can find advanced settings for tone control, microphone gain, recording monitoring, etc. Be careful with the recording monitor settings as they can cause an internal or external feedback loop.

As you look at these controls, make sure that the sliders are up, the balance is centered; and the inputs and outputs are unmuted. You can be assured that something *won't* work if you have any one of these sliders all the way down or that input muted. Again, your sound card may be different, but if it says "Select" instead of "Mute" you have to check the box instead of unchecking it.

Sound Card Interfaces

Configuring your sound card levels is one thing, but getting the audio to and from your radio is another. You also need to provide a way for your computer to switch your transceiver between receive and transmit. This is where the *sound card interface* comes into play.

Most amateurs use commercial sound card interfaces such as those made by MFJ Enterprises, West Mountain Radio, MicroHAM, MixW RigExpert and Tigertronics (see the "Software and Hardware" sidebar). These handy boxes take care of everything. They match the audio levels, isolate the audio lines and provide transmit/receive switching, usually with your computer's serial (COM) port.

Figure 6.3 shows a typical interface connection between a computer and a transceiver. Note that the transmit audio connects to the radio through the interface. By doing so, the interface can provide isolation.

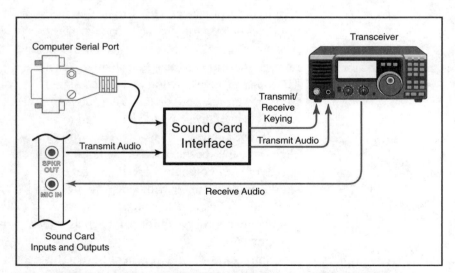

Figure 6.3—The sound card interface acts as the middleman between your radio and your computer.

Some interfaces also include a transmit audio adjustment, although this can also be accomplished at the computer, as we've already discussed.

Look closely at Figure 6.3 and you'll see that the receive audio cable is connected directly between the radio and the sound card input. Many interfaces use this approach, but others also pass the receive audio through the interface and provide isolation for this signal path as well.

It's important to mention that you may also be able to use the VOX (voice-operated switching) function on your transceiver to automatically switch from transmit to receive when it senses the transmit audio from your sound card. This approach completely removes the need for a T/R switching circuit, COM port and so on. The Tigertronics SignaLink interface uses this technique by incorporating its own VOX circuit, freeing the computer COM port for other uses.

The weakness of this technique is that it will cause your radio to transmit when it senses any audio from your computer—including miscellaneous beeps, music, etc.

Multimode Processors

Until sound card software appeared on the market, the most common HF digital interface was the *multimode processor*, or what some refer to as a *multimode TNC*. This device offers several digital communication modes in a single unit that usually sits alongside your radio.

The only amateur digital mode in use today that requires a multimode processor is PACTOR. To operate PACTOR II or III (the most common PACTOR modes), you'll need to

An SCS PACTOR controller. This device not only supports PACTOR, but several other digital modes as well.

purchase a "PTC" processor from SCS Corporation. See the "Software and Hardware" sidebar.

The Digital Transceiver

Just about any modern SSB transceiver will do double duty as an HF digital radio. The only concern is output power. When you transmit with modes that demand long transmit intervals, you're asking your rig to produce full output for several minutes at a time—maybe longer if you're a slow typist. This is known as a *100% duty cycle*. Most transceivers won't tolerate this kind of punishment; they're designed for the lower duty cycles of SSB or CW. When in doubt, reduce your output. Some manufacturers recommend a reduction of about 50%.

> *Just about any modern SSB transceiver will do double duty as an HF digital radio.*

For digital modes above 50 MHz, FM voice transceivers are ideal, especially for packet radio. For more advanced digital modes at VHF and UHF, special transceivers are required, such as those compatible with the D-Star digital protocol. You can learn more about D-Star on the Web at **www.icomamerica.com/amateur/dstar/**.

LET'S TRY PSK31!

PSK31 is by far the most popular digital mode, especially on the HF bands. You can find PSK31 conversations any time the band is open.

Jump onto the Web and download the PSK31 software you need, according to the type of computer system you are using. Since this is your first foray into digital communications, try one of the free programs mentioned in the "Software and Hardware" sidebar, such as *DigiPan*.

Once you have the software safely tucked away on your hard drive, install it and read the "Help" files. Every PSK31 program has different features--too many to cover in this chapter. Besides, the features will no doubt change substantially in the months and years that follow the publication of this book. The documentation that comes with your software is the best reference.

The Panoramic Approach

There is one feature that is common to every PSK31 application. It's important, so let's spend a little time talking about it.

One of the early bugaboos of PSK31 had to do with tuning. Most PSK31 programs required you to tune your radio carefully, preferably in 1-Hz increments. In the case of the original G3PLX software, for example, the narrow PSK31 signal would appear as a white trace on a thin waterfall display. Your goal was to bring the white trace directly into the center of the display, and then tweak a bit more until the phase indicator in the circle above the waterfall was more-or-less vertical (or in the shape of a flashing cross if you were tuning at QPSK signal). Regardless of the software, PSK31 tuning required practice. You had to learn to recognize the sight and sound of your target signal. With the weak warbling of PSK31, that wasn't always easy to do. And if your radio didn't tune in 1-Hz increments, the receiving task became even more difficult.

Nick Fedoseev, UT2UZ and Skip Teller, KH6TY, designed a solution and called it *DigiPan*. The "pan" in *DigiPan* stands for "panoramic"--a complete departure from the way most PSK31 programs worked. With DigiPan the idea is to eliminate tedious tuning by detecting and displaying not just one signal, but entire groups of signals.

If you are operating your transceiver in SSB without using narrow IF or audio-frequency filtering, the bandwidth of the receive audio that you're dumping to your sound card is about 3000 Hz. With a bandwidth of only about 31 Hz, many PSK31 signals can squeeze into that spectrum. Panoramic software continuously sweeps through the received audio from 100 to 3000 Hz and shows you the results in a large waterfall display that scrolls from top to bottom. What you see on your monitor are vertical lines of various colors that indicate every signal that the software can detect. Bright lines represent strong signals while faint lines indicate weaker signals.

The beauty of panoramic reception is that you do not have to tune your radio to monitor any of the signals you see in the waterfall. You simply move your mouse cursor to the signal of your choice and click. A cursor appears on the trace and the software begins displaying text. You can hop from one signal to another in less than a second merely by clicking your mouse. If you discover someone calling CQ and you want to answer, click on the transmit button and away you go--no radio adjustments necessary.

The Panoramic Downside

The weakness of panoramic reception is found in the fact that your receiver is running "wide open." The only IF filtering is typically your 1.8 or 2.5 kHz SSB filter. The automatic gain control, or AGC, circuit in your receiver is acting on everything within that bandwidth, working hard to raise or lower the overall gain according to the overall signal strength. That's fine if all the PSK31 signals are approximately the same strength, but if a very strong signal appears within the bandwidth, the AGC will *reduce* the gain to compensate.

The result will be that many of the signals in the waterfall display will suddenly vanish, or become very weak, as the AGC drops the receiver gain. In cases where an extremely strong signal appears, *all* signals except the rock crusher may disappear completely.

Some PSK31 operators are using transceivers that allow them to enable 500-Hz CW filters when the rig is in the SSB mode. They'll switch in the filters and sacrifice their wide waterfall displays to remove strong signals nearby.

Tuning PSK31

Most of the PSK31 activity can be found on 20 meters around **14070 kHz**, but you'll also find PSK31 on the following frequencies:

3580 kHz
7070 kHz
10140 kHz
21070 kHz
28120 kHz

Start by putting your radio in either the SSB mode, either USB or LSB. Park your radio on the frequency of your choice and boot up your PSK31 software. *Do not touch your rig's VFO again.*

Why not?

The frequency your transceiver displays in the SSB mode is the *suppressed carrier frequency*. If you've selected upper sideband on your radio (USB), your receiver range is everything from the suppressed carrier frequency to about 2 or 3 kHz above it. If you select lower sideband (LSB), your receiver range extends 2 or 3 kHz *below* the suppressed carrier frequency.

For example, let's say that you've dialed your transceiver to 14070 kHz and you have the

> **Park your radio on the frequency of your choice and boot up your PSK31 software. Do not touch your rig's VFO again.**

radio in USB mode. Assuming that you don't have a narrow IF filter enabled, your receiver is picking up everything from 14070 to about 14073 kHz. The PSK31 software will display all signals within that range. Just place your mouse cursor on one of the vertical signal lines and right click. The software selects that signal without the need for you to tune your radio and begins displaying the text. When you transmit, the PSK31 software generates a tone that corresponds to the tone of the signal you selected. When that tone is applied to your radio, it creates an RF signal on the correct frequency. That's all there is to it!

PSK31 signals have a distinctive sound unlike any digital mode you've heard on the ham bands. PSK31 signals *warble*.

Using waterfall displays you may often see (and copy) PSK31 signals that you cannot otherwise hear. It is not at all uncommon to see several strong signals (the audible ones) interspersed with wispy blue ghosts of very weak "silent" signals. Click on a few of these ghosts and you may be rewarded with text (not error-free, but good enough to understand what is being discussed).

Does the Sideband Matter?

With BPSK, the most popular PSK31 mode by far, the answer is "no." You can have your transceiver in upper or lower sideband and work any signals that appear in your software display.

QPSK is another matter. Sideband selection is critical for QPSK. Most QPSK operators choose upper sideband, although lower sideband would work just as well. The point is that both stations must be using the same sideband, whether it's upper or lower.

A PSK31 Conversation

Click on the lines in the waterfall display and you may see something like this...

CQ CQ CQ CQ CQ DE WB8IMY WB8IMY WB8IMY K

That's a PSK31 operator calling CQ and hoping to make contact. If you've set up your software properly, you're ready to reply. Just put your PSK31 program into the transmit mode and start typing...

WB8IMY WB8IMY DE N6ATQ N6ATQ K

"K" means "over to you," just like it does on CW. Now put your software back into the receive mode. With luck you'll be heard and the conversation will begin!

Hello! I'm seeing perfect text on my screen, but I can barely hear your signal. PSK31 is amazing! N6ATQ DE WB8IMY K

Some PSK31 programs and processor software offer type-ahead buffers, which allow you to compose your response "off line" while you are reading the incoming text from the other station.

Almost all programs offer pre-programmed messages known as macros. A macro allows you create a sequence of actions that will take place whenever you press a particular key on your keyboard. Every program uses different macro symbols, but here is a typical example...

<TX>
CQ CQ CQ CQ CQ CQ DE WB8IMY WB8IMY
CQ CQ CQ CQ CQ CQ DE WB8IMY WB8IMY K
<RX>

Whenever you press the assigned key, this macro will place your radio in the transmit mode <TX>, send two lines of CQ text, then switch your radio back to the receive mode <RX>. CQ macros are common because they streamline the process of calling a CQ.

Depending on your software, you may be able to set up more elaborate macros, like this one...

```
<TX>
<HISCALL> DE WB8IMY. Thank you for the call. Your signal report is
<SIGNALRPT>. My QTH is Wallingford, CT, USA. Back to you. <HISCALL> DE
WB8IMY K
<RX>
```

The macro starts by placing the transceiver into the transmit mode <TX>.
<HISCALL> inserts the call sign of the calling station (assuming that you've entered it
into the software's log) and then begins sending text. The signal report <SIGNALRPT> is
inserted automatically. Finally, <HISCALL> appears again as you turn it back to him and
return your radio to the receive mode <RX>.

Be careful about relying too much on macros. They are convenient, but they can also
cause confusion if used improperly. In addition, some hams create macros that contain
lengthy descriptions of their station equipment, families, etc. It's tempting to tap that macro
key and sit back while your computer sends more information than the other person really
wants to know!

CHASING CONTACTS WITH RTTY

If there is a contest going on, or if a rare DX station is on the air, chances are good that
RTTY will be the mode of the day. If you're set up for PSK31, you're set up for RTTY, too.
Just change programs and give it a try.

How RTTY Works

Each character in the Baudot RTTY code is composed of five bits. In amateur RTTY
communication a "1" bit is usually represented by a 2125-Hz tone and is known as a *mark*.
A "0" bit is represented by a 2295-Hz tone called a space. There is also a start pulse at the
beginning of the bit string and a stop pulse at the end. The data is commonly sent at a rate
of 60 WPM, or 45 baud.

The difference between the
mark and space tones is known
as the shift. Grab your calculator
and do a bit of subtraction:

$$2295 \text{ Hz} - 2125 \text{ Hz} = 170 \text{ Hz}$$

> *If there is a contest going on, or if a rare
> DX station is on the air, chances are good
> that RTTY will be the mode of the day.*

In the answer shown above, 170 Hz is the difference or shift between the mark and
space frequencies. The Amateur Radio RTTY standard is to use either a 170- or 200-Hz
shift.

When setting up your RTTY software, you may be asked to choose the shift. For
Amateur Radio use, select 170 Hz.

Tuning and Decoding the RTTY Signal

Every RTTY decoder, whether it works in software or hardware, incorporates a set
of mark and space audio filters. You can think of these filters as dual windows that only
open for tones that are at the correct mark and space frequencies, and separated by the
proper shift. The mark and space filtering circuitry detects and decodes the tones into
digital 1s and 0s, which is exactly what your computer or processor needs to provide text
on your screen. The more sensitive and selective your mark/space detectors, the better your
RTTY performance, especially as it involves your ability to copy weak signals through
interference.

It's easy to understand why tuning a RTTY signal is so critical—and why a good tuning indicator is one of your best HF digital tools. Whenever you stumble upon a RTTY signal, you must quickly tune your receiver until its mark and space tones fall within the "skirts" of the filters and are detected. It's possible to do this by ear once you become accustomed to the sound of a properly tuned RTTY transmission, but few of us have the necessary patience. Instead we rely on visual indicators to guide us. Fortunately, every RTTY program has a tuning indicator of some kind.

CONVERSING WITH RTTY

As with so many aspects of Amateur Radio, begin by listening. Tune between 14.070 and 14.099 MHz and listen for the long, continuous *blee-blee-blee-blee* signals of RTTY. (If you hear warbling, it isn't RTTY!)

Make sure your transceiver is set for lower sideband (LSB). That is the RTTY convention. The exception is when you are operating your rig in the "FSK" mode (sometimes labeled "DATA" or "RTTY").

As you tune the RTTY signal, watch your tuning indicator. Tune slowly until you see that the mark and space tones are being decoded. At this point you should see letters marching across your screen. Notice how the conversation flows just like a voice or CW ragchew.

KF6I DE WB8IMY . . . YES, I HEARD FROM SAM JUST YESTERDAY. HE SAID THAT HIS TOWER PROJECT WAS ALMOST FINISHED. KF6I DE WB8IMY K

If you want to call CQ, the procedure is simple. Some programs have "canned" CQ messages that you can customize with your call sign. Others allow you to type off the air, filling a buffer with your CQ and storing it temporarily until you're ready to go.

Let's assume that you have a CQ stored in your buffer right now. Press the key that puts your rig in the transmit mode. Now tap the key that spills the contents of your buffer into the modem. If you're monitoring your own signal you'll hear the delightful chatter of RTTY and see something like this on your screen:

CQ CQ CQ CQ CQ CQ WB8IMY WB8IMY WB8IMY
CQ CQ CQ CQ CQ CQ WB8IMY WB8IMY WB8IMY K K

Now jump back to receive. That's all there is to it!

Notice how my CQ is short and to the point. Remember that RTTY lacks error detection. If you want to make certain that the other station copied what you sent, it helps to repeat it. You'll see this often in contest exchanges. For example:

N6ATQ DE N1RL . . . UR 549 549. STATE IS CT CT. DE N1RL K

On the other hand, if you know that the other station is copying you well, there is no need to repeat information.

KEEP EXPLORING

Once you become familiar with PSK31 and RTTY, expand your horizons. There is much more for you to learn and enjoy!

Try one of the visual modes such as SSTV. Just download some SSTV software and tune between 14230 and 14240 kHz, particularly on weekends. You'll hear some strange-sounding signals that are probably SSTV. Fire up your program and eavesdrop for a while, viewing the images stations are sending to each other. There is more about SSTV on the ARRL Web at **www.arrl.org/tis/info/sstv.html**.

Try some of the lesser-used digital modes such as Olivia, MFSK16 and Hellschreiber.

Try some of the lesser-used digital modes such as Olivia, MFSK16 and Hellschreiber. You'll occasionally hear these signals scattered between 14080 and 14110 kHz, and on other bands. Olivia, in particular, has remarkable performance under weak signal conditions and it is gaining popularity as a result.

Above 50 MHz, try bouncing digital signals off the trails of meteors as they burn in the upper atmosphere. As wild as this sounds, hams do it all the time using the free *WSJT* software package. See Chapter 9.

Tune your FM transceiver to 144.39 MHz and listen. Do you hear bursts of digital noise? If so, you're probably hearing the sounds of APRS, the Automatic Position Reporting System. APRS stations send position reports at regular intervals using packet radio. The APRS software decodes the position reports and displays the results as icons on a computer-generated map. Many of these APRS stations are mobile and have Global Positioning System (GPS) receivers attached to packet interfaces, which are in turn wired to FM transceivers. As their positions change, so do the map symbols!

You can learn more about APRS on the ARRL Web at **www.arrl.org/tis/info/HTML/aprs/**. Free APRS software for *Windows* is available as well. See *UI-View* 16 in the "Software and Hardware" sidebar.

7 Chasing Contacts and Wallpaper

You may not think of Amateur Radio in terms of "competition," but thousands of hams see it exactly that way. For them, ham radio is almost a sport. They compete against their fellow amateurs in contests of skill, or they simply compete against themselves and the vagaries of propagation. Either way, it is the thrill of hunt that gets their blood pounding.

Kristin, KCØINX, and Kathryn, KSØP, compete in the ARRL Sweepstakes contest.

CONTESTS

Amateur Radio contesting has a deceptively simple goal: to contact as many stations as possible during the contest period. Of course, you know that it really isn't that straightforward. Basketball is about more than putting a ball through a hoop, and ham contesting is about more than making contacts.

Every contest has specific rules (just like every athletic sport). For example…

■ Only certain bands may be used.

■ The contest only takes places between certain times, and on certain dates. Some contests also require "off times" when you are forced to leave the air.

■ An exchange of information is necessary during each contact. You may be required to send and receive a serial number, location, name or even a person's age.

■ Only certain operating configurations can be used. You may have to choose a "class" of operation such as a single operator using low power. (See the sidebar "Multi *What?*")

There is a contest of one type or another almost every weekend. The most popular contests are sponsored by the ARRL, but there are many others. Some competitions, such as the ARRL Sweepstakes, draw large numbers of hams onto the airwaves. Other contests are smaller with only limited participation.

Contests take place primarily on the HF bands, with the exceptions of 60, 30, 17 and 12 meters. Contest sponsors have agreed to keep these bands off limits from competition. There are also contests on the VHF, UHF and microwave bands. A list of ARRL contests is shown in **Table 7-1**.

The best way to keep track of contest activity is by becoming an

> *There is a contest of one type or another almost every weekend.*

Table 7-1

ARRL Contests

For specific dates and rules, see the ARRL Web at **www.arrl.org**, or *QST* magazine.

Month	Contest	Comments
January	RTTY Roundup	Digital only
	VHF Sweepstakes	
February	International DX—CW	CW only
March	International DX—phone	SSB only
June	VHF QSO Party	
July	IARU Championships	HF bands only
August	UHF Contest	
	10 GHz and Up	
September	VHF QSO Party	
	10 GHz and Up	
November	Sweepstakes—CW	CW only
	Sweepstakes—phone	SSB only
December	10 Meter	
	160 Meter	CW only

ARRL member. As a member you'll receive *QST* magazine each month. In every issue you'll find "Contest Corral," a comprehensive list of upcoming contests, both ARRL and otherwise. You will also find a contest calendar on the ARRL Web at **www.arrl.org**. The ARRL even offers an e-mail newsletter called the *Contester's Rate Sheet* at **www.arrl.org/contests/rate-sheet/**. Last, but certainly not least, the ARRL publishes a bimonthly magazine devoted exclusively to the joys of contesting: the *National Contest Journal* (*NCJ*). Learn more at **www.arrl.org/ncj/**.

Contest Logging

The Federal Communications Commission does not require hams to keep station logs with records of every contact, but contest sponsors *do*. Your log is your contest entry; without it, your score won't even be considered.

You can keep a contest log on paper and submit the paper log at the end of the competition. Most contesters, however, do their logging by computer. Frankly, computer logging is *much* easier than paper logging. The computer keeps track of the time, score and much more.

Your computer will also help you avoid the dreaded *dupe*—the duplicate contact. Depending on the rules of the contest, you may only be allowed to contact a particular station once on a given band. For instance…

WB8IMY contacts K1RO on 40 meters at 0100 UTC. Score = 1 point

WB8IMY contacts K1RO on 20 meters at 0300 UTC. Score = 1 point

WB8IMY contacts K1RO on 40 meters at 0530 UTC. Score = *zero!* This contact is a dupe of the previous 40-meter contact at 0100.

Contest software will alert you to possible dupes before you waste time making the contact. If you hear someone calling "CQ Contest" and you type their call sign into the log, the software will instantly check and make sure that a contact with the station is "legal" under the rules. If working that station constitutes a dupe, you'll know right away.

Contest software also makes it easy to submit your log after the contest is over. The contest sponsors supply e-mail addresses for you to send your log, along with a brief description of your station and entry classification (known as a *summary sheet*).

Multi *What?*

For every contest you must choose your operating classification. The classifications depend on the rules of the contest. Here are some typical examples…

● Single Operator, Low Power (SOLP): One person operating one transceiver at a power level of 100 W or less.

● Single Operator, High Power (SOHP): One person operating one transceiver at a power level greater than 100 W.

● Single Operator, Two Radios (SO2R): One person operating two transceivers simultaneously. A real juggling act; not for the faint hearted!

● Single Operator, Single Band: One person operating one transceiver, but the station must stay on a single band throughout the contest.

● Multioperator: Several operators at a single station, but only one radio can be on the air at a time.

● Multi-Multi: Several operators at a single station with several radios on the air at the same time. This is the ultimate contest party!

● Rover: This is a VHF/UHF/microwave classification. A Rover operates from a location in one *grid square* (see Chapter 9), then moves and operates from another grid square.

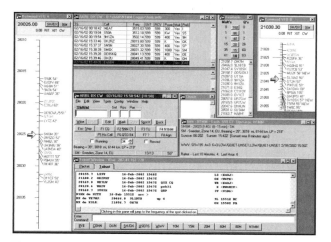

The *N1MM Logger* program.

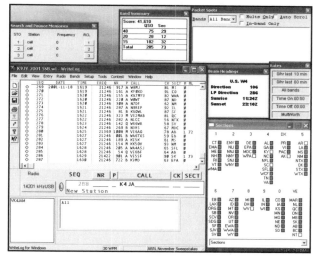

WriteLog in use during the ARRL Sweepstakes contest.

Although preparing these documents for e-mailing may sound like a hassle, it isn't. The contest software will create these files for you with a few clicks of your mouse.

Most contest logging software is written for Microsoft *Windows*. Some of the popular titles include…

N1MM Logger: **pages.cthome.net/n1mm/**

WriteLog: **www.writelog.com/**

TRLog: **www.trlog.com/**

There is also an excellent program for Macintosh computers known as *MacLoggerDX*. You'll find it at **www.dogparksoftware.com/MacLoggerDX.html**.

The Importance of UTC

"What time is it?"

Sounds like an easy question, doesn't it? If you're talking to someone in your own time zone, the answer is indeed easy since both of your clocks agree. But what if the person you are talking to lives many time zones away? Now the answer becomes more complicated. For instance, if it is 11PM Tuesday in New York, it is noon *Wednesday* in Tokyo, Japan!

Imagine the confusion of trying to reconcile all the time differences that exist in the world. Long ago people decided to minimize the headaches by using one time standard as a reference for the entire globe. This reference used to be established by clocks at an observatory in Greenwich, England and was known as Greenwich Mean Time. It has since been renamed Universal Time Coordinated, or UTC.

Hams use UTC for all timekeeping because everyone in the world easily understands it. We also use the 24-hour time format. Rather than saying "4PM," for instance, we say "1600."

If you want to meet your friend in Tokyo on the air, you can say, "I'll be on 15 meters at 2300 UTC." If you both know how to convert from UTC to your local time, you'll know immediately when you should be at your radios.

These are the UTC conversions for hams in US time zones. Subtract these hours from UTC to determine your local time.

Eastern	Central	Mountain	Pacific	Alaska	Hawaii
–5 hours	–6 hours	–7 hours	–8 hours	–9 hours	–11 hours

If the area where you live observes Daylight Saving Time, subtract one hour *less* than the number shown above.

When considering UTC, don't forget the date. For example, let's say that you live in Denver, Colorado, which is usually 7 hours behind UTC. It is Friday evening and you just completed a contact at 2000 local time. Translating to UTC, 2000 local time becomes 0300. Look carefully at the UTC time, though. It is after midnight, isn't it? The contact may have occurred Friday evening by your clock and calendar, but it was *Saturday morning* UTC! When you record the date of the contact, you need to show it as Saturday, not Friday.

Running vs Searching Pouncing

You'll often hear contesters speak of *running*. This means finding a clear frequency and calling "CQ contest" for long periods of time, logging everyone who answers. Running is an effective contest strategy if your station has a big signal that many can hear. You'll be like a contest beacon, drawing the multitudes to you.

On the other hand, if you have a smaller signal profile you might want to consider *searching and pouncing*, or S&P. Just like the term implies, this involves tuning through the frequencies, looking for the running stations and contacting any you can find. Even though your signal may be weak, the runners will make special efforts to pick you out of the noise because they need the points your contacts will give them.

A typical SSB contest contact between a runner (K1ZZ) and an S&P operator (N5RL) looks something like this…

CQ contest, CQ contest from K1ZZ, Kilowatt-One-Zulu-Zulu. Contest!

Rules Are Meant to be Followed

Each contest has its own set of rules. Some are simple and others are elaborate. Your key to success is understanding the rules and following them explicitly. Here is an example of a typical contest rule set. These belong to the ARRL 10-Meter Contest…

1. Object: For Amateurs worldwide to exchange QSO information with as many stations as possible on the 10-meter band.

2. Date and Contest Period: Second full weekend of December. Starts 0000 UTC Saturday; ends 2359 UTC Sunday.

2.1. All stations operate no more than 36 hours out of the 48-hour period.

2.2. Listening time counts as operating time.

3. Entry Categories:

3.1. Single Operator: (9 categories)

3.1.1. QRP.

3.1.1.1. Mixed Mode (Phone and CW).

3.1.1.2. Phone only.

3.1.1.3. CW only.

3.1.2. Low Power.

3.1.2.1. Mixed Mode (Phone and CW).

3.1.2.2. Phone only.

3.1.2.3. CW only.

3.1.3. High Power.

3.1.3.1. Mixed Mode (Phone and CW).

3.1.3.2. Phone only.

3.1.3.3. CW only.

3.2. Multioperator, Single Transmitter, mixed mode (only).

3.2.1. Includes single operators using packet or spotting assistance.

4. Contest Exchange:

4.1. W/VE stations (including Hawaii and Alaska) send signal report and state or province (District of Columbia stations send signal report and DC).

4.1.1. Novice and Technician Plus stations sign /N or /T on CW. If used, you must indicate /N or /T on your summary sheet.

4.2. DX stations (including KH2, KP4, etc) transmit signal report and sequential serial number starting with 001.

4.3. Maritime mobile stations send signal report and ITU Region (R1, R2 or R3).

5. Scoring:

5.1. QSO points:

5.1.1. Two points for each complete two-way phone QSO.

5.1.2. Four points for each two-way CW QSO.

5.1.3. Eight points for CW QSOs with US Novice or Technician Plus stations signing /N or /T (28.1 to 28.3 MHz only).

5.2. Multipliers: (per mode, phone and CW).

5.2.1. Each US state and the District of Columbia.

5.2.2. Canada [NB (VE1, 9), NS (VE1), QC (VE2), ON (VE3), MB (VE4), SK (VE5), AB (VE6), BC (VE7), NWT (VE8), NF, (VO1), LB (VO2)], YT (VY1), PEI (VY2) NU (VYØ).

5.2.3. DXCC countries (except US and Canada). KH6 and KL7 participate and count as US states and send HI or AK as that part of their exchange.

5.2.4. ITU regions (maritime mobiles only).

5.3. Final Score: Multiply QSO points by total multipliers (the sum of states/VE provinces/DXCC countries/ITU regions per mode). Example: KA1RWY works 2245 stations including 1305 phone QSOs, 930 non-Novice CW QSOs, 10 Novice CW QSOs, for a total of 6410 QSO points. She works 49 states, 10 Canadian call areas, 23 DXCC entities and a maritime mobile station in Region 2 on phone and 30 states, 8 Canadian call areas, and 19 DXCC countries on CW for a total multiplier of 140. Final score = 6410 (QSO points) × 140 (multiplier) = 897,400 points.

6. Miscellaneous:

6.1. Single operator mixed-mode and multioperator stations may work stations once on CW and once on SSB.

6.2. Your call sign must indicate your DXCC country if competing as DX. (N6TR in Oregon does not send N6TR/7, but K1NO in Puerto Rico must send K1NO/KP4).

6.3. All entrants may transmit only one signal on the air at any given time.

6.4. All CW contacts must take place below 28.3 MHz.

N5RL (N5RL answers)

N5RL copy 599 Connecticut. QSL? (K1ZZ acknowledges and gives the required exchange. In this case, it is his signal report and state)

K1ZZ QSL. Copy 599 Texas. (N5RL acknowledges and gives his signal report and state.)

QSL Texas. Thanks for the contact. K1ZZ QRZ! (K1ZZ thanks him for the contact and says "QRZ" to ask if any other stations wish to call.)

CW and digital contacts take place in much the same way. CW contesters tend to send and receive at high speeds, but they will usually slow down for slower operators. Digital operators use software that allows most of the contest exchange to be sent automatically by pressing single keyboard keys.

Tips from the Winners

The hams who do consistently well in contests have a number of things in common: They all follow certain habits that work to enhance their performance and their score. Borrowing from their playbooks, here are the top tips…

1. Read the rules well before the contest and make sure you understand them. See the sidebar, "Rules Are Meant to be Followed" for a typical example.

2. Check all your equipment (including software) a few days before the contest begins. Make sure everything is operating perfectly.

3. Understand the basics of propagation (see Chapter 3) and plan your contest strategy accordingly. Try to obtain a propagation forecast for the contest weekend. These are usually posted on the ARRL Web at **www.arrl.org**.

4. Make plans for rest and nourishment. Have food and drink on hand. Take breaks every couple of hours to stretch your legs and clear your mind.

It Isn't All About "Winning"

Even though contest competition can be intense, it isn't always about winning. You may never win the top slot in a contest, but you'll definitely enjoy the competition and the camaraderie. "Multi" operations are particularly rewarding because you're contesting with a team rather than by yourself.

When the contest sponsors post the results, you'll be able to look at your score with pride because you know how much effort it took to get there. Maybe

---

Contest Behavior

In the rush of competition it is easy to forget that you're sharing the airwaves with other hams who have no interest whatsoever in contesting. Look at the situation from their point of view. Maybe they wanted to get on the air for casual weekend chats with friends. They turn on their radios and hear "CQ Contest" blasting from their speakers on every available frequency. How would you feel if you were in their shoes?

If you're operating a contest, use a large measure of courtesy and consideration. Don't choose a running frequency that is adjacent to an ongoing conversation. Your signal is almost certain to interfere.

If you are asked to move because someone wants to use the frequency, or because you are causing interference, do the right thing. No, you are not legally obligated to move. You are within your rights to stay right where you are. But if you dig in your heels and refuse, consider the "message" you are sending to the non-contest community. No frequency is sacred, even in a contest. If someone politely asks you to move, politely acknowledge and find another frequency.

> **DX does indeed involve distance, but it also involves difficulty and scarcity.**

your score was better than what you earned last year, or perhaps you'll indulge in the guilty pleasure of seeing that you beat another ham that you know personally. (Contesting is friendly competition, after all!) Either way, you'll have fun and polish your operating skills at the same time. That's not a bad way to spend a weekend!

IN SEARCH OF DX

We tend to think of "DX" in the literal sense—a station operating from a distant location. While that may be true in some circumstances, the definition of DX is really in the eye of the beholder.

For instance, what sort of distance defines DX? If you're operating on the HF bands, that distance may be measured in thousands of miles. For a microwave operator, however, DX may mean a contact over a hundred miles or so.

DX does indeed involve distance, but it also involves difficulty and scarcity. Take the little island of Desecho in the Caribbean. Desecho is near Puerto Rico, which makes it practically next door for signals on the HF bands. If distance is your only yardstick, Desecho doesn't qualify as DX as far as HF operators in the United States are concerned.

But Desecho is uninhabited. No hams live on Desecho and the island is difficult to reach because of a tangle of regulations that prohibit most visitors. Therefore, it is a rare day when a ham journeys to the island and puts a signal on the air. Such an event is big news in the DX community as thousands rush to their radios in the hope of making contact with Desecho before the hams pull down their antennas and leave the island.

So is Desecho Island DX? You bet it is! It isn't distance that makes Desecho such a big DX prize—it is because the island is rarely on the air thanks to the difficulty of getting there.

A large segment of Amateur Radio consists of hams who devote their time to making contacts with stations in DX locations, many of them quite rare. We call these individuals *DXers*. They keep track of all the DX contacts they've made because these confirmed contacts count as credit for coveted awards. Hams affectionately refer to award certificates as *wallpaper*.

The most coveted DX wallpaper in the world just happens to be sponsored by the ARRL. It is the DX Century Club award, or DXCC for short.

DXCC

The criteria for earning a basic DXCC award seems simple on the surface. You must make confirmed contacts with 100 DXCC *entities* using voice, CW or digital, or a mixture of modes. If you live in the United States, you must also be a member of the ARRL to apply for the award.

So what is a DXCC entity?

1. A country that is a

The most sought-after operating award in the world: the DX Century Club.

member of the United Nations, or an area that has been assigned a call sign prefix by the International Telecommunication Union (ITU). This definition includes those places that we commonly think of a "countries" such as Germany, Saudi Arabia and so on.

2. Any piece of land that is separated from a "parent" DXCC entity at a specific distance according to the nature of the separation. (The DXCC rules describe this qualification in mind-numbing detail, so we won't discuss it here!) Desecho Island qualifies as a DXCC entity under this rule.

Don't worry about having to learn the rules for DXCC entities. All you need to know are which entities are "legal" for DXCC credit. Just get on the Web and go to **www.arrl.org/awards/dxcc/**. That's where you'll find the complete DXCC entities list. You can also purchase a printed version of the list from the ARRL.

Your journey on the road to a basic DXCC award is much like riding a bicycle up a gradual incline. At first it is easy. Make a confirmed contact with another US station and you've earned your first DXCC entity credit. (Yes, the US is a DXCC entity.) Confirm a conversation with a ham in Canada and you've earned another. There are lots of hams on the air in Great Britain. Work one of them and you've earned your third DXCC credit.

Bagging entities with large ham populations is simple, but as you run out of those entities you must turn your attention to places in the world that harbor few hams, or perhaps none at all. Now the DXCC hill becomes much steeper. You soon find yourself searching eagerly for those "rare" entities that you need to put yourself over the 100-credit mark.

Confirming 100 entities and earning a basic DXCC award is only the beginning. Believe it or not, there are well over *300* entities on the DXCC list. You earn endorsement stickers as you climb the hill from 100 to 125, 150, 175 and upward. Some entities are so rare that hams are only able to put them on the air once per *decade*. Despite this, there are a few persistent and patient souls who have actually managed to work them all!

And then there are the various permutations of the DXCC award. There are hams who only chase DX on a particular mode, such as CW, so they seek CW-only awards. There are hams who earn DXCC awards on individual bands (100 entities or more on 20 meters, for instance). There is a Five-Band DXCC. That means that you have made confirmed contacts with 100 entities on *each* of the five qualifying HF bands: 80, 40, 20, 15 and 10 meters. Finally, there is the *crème de la crème:* the DXCC Challenge Award for contacting a combined total of 1000 DXCC entities.

Confirmation

You'll notice that we keep making references to "confirmed" contacts. It isn't enough to simply say that you've contacted a station. To earn DXCC, or any other award, you must be able to *prove* your contacts.

Until recent times, the only acceptable proof was a postcard known as a *QSL*. QSLs are often highly personalized with art or photographs.

IZ8GBH proudly displays a long-awaited QSL that just arrived at his QSL bureau in Italy.

Here's a QSL card from Antarctica—R1ANF!

A DXpedition QSL card from Wake Island.

Who couldn't love a QSL card like this?

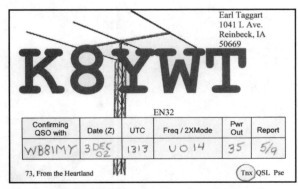

Your own QSL card can be simple, yet attractive, like this one.

As a new ham you'll no doubt want to order a supply of your own, so check the Web for QSL printers, or the advertising pages of *QST* magazine. The cards are printed with your name, call sign, address and any other information you wish. They also include blank spaces for you to add contact information.

If you want a QSL confirmation for a contact, you must send your card to the other station and request his card in return. Your QSL must include the following details:

1. The call sign of the station you contacted
2. The date of the contact (in UTC)
3. The time the contact began (also in UTC)
4. The frequency
5. The mode (CW, SSB, PSK31, etc)

Direct or Via the Bureau?

When you have the card filled out and are ready to mail, you must make a decision. Will you send the card directly through the mail, or via the QSL Bureau system?

If you send your card directly, you'll need to look up the station's postal address. Fortunately, this is easily done at Web sites such as **www.qrz.com**. Some DX stations use QSL managers to handle the flood of cards. These are hams who have volunteered to act as agents for the DX station. If you discover that a QSL manager exists, send your card to the manager's address, not the DX station's address.

Once you have the address, the next step is to put your QSL into an airmail envelope with (1) a self-addressed return envelope and (2) two dollars in cash (two 1-dollar bills) to pay for the return postage. The money is necessary because the ham on the other end will need to buy stamps to send his card back to you; US stamps won't work. (The exception is if the DX station has a QSL manager in the United States. If that is the case, you only need to include a self-addressed stamped envelope since you won't be sending the card overseas.)

You've probably guessed by now that direct QSL exchanges can become expensive over time. Considering the postage to get the card to the DX station, and the $2 for return postage, each QSL can cost $3 or more. Multiply that figure times 100 to earn a basic DXCC award and you're looking at an investment of more than $300.

The alternative is the QSL Bureau system—a network of miniature QSL post offices. There are Amateur Radio organizations in almost every nation on Earth and most maintain a QSL bureau—including the

ARRL. If you are an ARRL member, you can send your cards to the ARRL Outgoing QSL Service. The ARRL will, in turn, send your cards to the bureaus in the countries you've contacted. The bureau in the destination country will then send your card to the DX station. The cost to send QSLs in this fashion is far less than sending direct. You can learn more at **www.arrl.org/qsl/**. Only ARRL members can use the outgoing service.

Incoming QSL cards go to ARRL-sponsored bureaus in various US call sign districts. To receive these cards you need to keep a self-addressed stamped envelope on file at the bureau that serves your call district. Be careful, though. The bureau that will receive cards for you may not be the bureau in the area where you live. For instance, WB8IMY lives in Connecticut (in the 1st call district), but because his call sign is from the 8th district he must keep an envelope at the 8th district incoming QSL bureau, *not* at the 1st district bureau where he lives. To see a list of incoming QSL bureaus, go to **www.arrl.org/qsl/ qslin.html**.

So which is better: direct or via the bureau? The advantage of direct QSLing is speed. If the DX station or manager is on the ball, you'll have your confirming QSL in a month or so. On the other hand, direct QSLing can be expensive. The advantage of the bureau system is that it is very inexpensive—much less than the cost of sending direct. The disadvantage is that the bureau system can be very slow. It is not uncommon to wait for a year or longer to receive a card through the bureau.

The Electronic QSL Alternative

In the age of the Internet, it seemed as though there should be an affordable electronic alternative to the traditional QSL process. The problem is that electronic information can be easily faked and false confirmations would quickly undermine the integrity of the awards. Was it possible to create a secure system that would allow for trustworthy

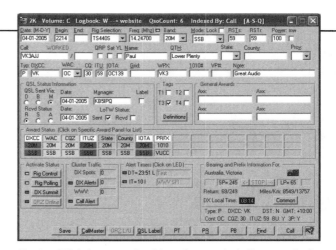

Prolog2K deluxe contact logging and station control software.

Logging Software

Computers have made station log keeping infinitely easier than it used to be. Instead of yellowing notebooks of torn paper, we now have software that will meticulously track and sort all of our contacts.

The first computer logging programs were little more than simple databases. They held records for each contact and not much more.

Modern logging programs are light years removed from their primitive ancestors. Even the least expensive programs can hold all your contact records, sort them any way you wish and tabulate your progress toward various awards such as DXCC and WAS.

More sophisticated applications do everything but make the contact for you! They will print labels for your QSL cards, connect to DX Clusters and alert you when a needed station appears on the air, connect to your transceiver and read its frequency, rotate your antennas, upload your data to Logbook of The World, look up station addresses and much more.

Prices range from about $20 to $100, depending on the complexity of the software. If you're looking for logging software, your best options are to check the Web as well as the advertising pages of *QST* magazine.

Just remember to back up your data regularly. You never know when a computer failure will erase your log completely. Safety is the one remaining advantage of paper logging!

> *Logbook of The World, or LoTW, is a huge repository of electronic contact logs from stations throughout the world.*

electronic QSLs to be exchanged via the Internet?

The ARRL worked on the problem and finally conceived an answer. They called it *Logbook of The World*.

Logbook of The World, or *LoTW*, is a huge repository of electronic contact logs from stations throughout the world. In order to use LoTW, one must (1) download the LoTW software; (2) submit a request for a digital certificate from ARRL; (3) use that certificate to "sign" your electronic log file; (4) submit the signed file to LoTW via e-mail or Web site upload; and (5) receive a confirmation from LoTW acknowledging receipt of the log data. You'll find complete details at **https://p1k.arrl.org/lotw/getstart**. Any ham can join LoTW and upload logs (uploading is free), but among US hams only ARRL members can redeem confirmation credits for awards such as DXCC.

When you upload a log, LoTW instantly compares it to every other log in the system. If it finds a match for a contact, you receive instant QSL credit! As you make more contacts, you can upload more logs and possibly receive credit for some of those contacts as well. It all depends on whether the stations you've contacted also participate in LoTW.

Electronic QSLing will never replace the thrill of receiving a personalized QSL card in your mailbox. But for hams who simply want the confirmation credits and don't care about cards, LoTW is a fantastic alternative. Many hams use both!

Worked All States

As we've already discussed, DX doesn't necessarily mean a station in a far away land. For hams chasing the ARRL Worked All States (WAS) award, the DX targets are the US states themselves.

> *The goal of the basic WAS is to make at least one confirmed contact with a ham in each of the 50 US states.*

The goal of the basic WAS is to make at least one confirmed contact with a ham in each of the 50 US states. As with DXCC, this seems easy at first. There are huge numbers of hams in California, for instance, so snagging your California contact shouldn't present much of a problem. On the other hand, amateurs in states such as North and South Dakota are few and far between.

You'll need to search the bands for contacts with the states you need. Here is a valuable tip, though: Contests are great tools for earning your Worked All States. The ARRL Sweepstakes, for example, can give you contacts in all 50 states in a single weekend!

As with the DXCC award, you must prove your contacts with QSLs—either paper or electronic (through Logbook of The World). If you want to pursue paper QSLs, it is generally easier to get return cards from hams in the States. Just make

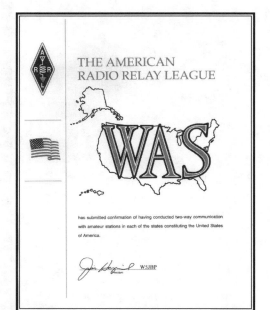

The Worked All States award.

sure to include a self-addressed stamped envelope and you'll greatly increase your chances of success. All domestic QSLs must be sent directly; you can't use the bureau system to exchange US QSLs. You can learn more about the Worked All States award at **www.arrl. org/awards/was/**.

Worked All Continents

This award is sponsored by the International Amateur Radio Union (IARU) and issued by the ARRL. You earn your Worked All Continents by making contacts with stations in all six continents (North America, South America, Oceania, Asia, Europe and Africa) on a variety of different bands and modes. This award is a good starting point on the way to your DX Century Club. More details are available at **www.iaru.org/wac/**.

Worked All Continents.

VHF/UHF Century Club

VUCC, as it is called, is the king of DX awards in the world above 50 MHz. The goal isn't to contact stations in DXCC entities. Instead, VUCC enthusiasts pursue *grid squares*.

Grid squares are more properly known as *Maidenhead grid squares*. They are based on an idea originally put forth by British amateur John Morris, G4ANB, and later adopted by a group of VHF enthusiasts meeting in Maidenhead, England in 1980.

The most coveted award above 50 MHz: VHF/UHF Century Club.

The Maidenhead system divides the entire world into small rectangles that are two degrees longitude by one-degree latitude. These grid squares are a convenient means of determining the approximate location of a station across a broad swath of land (such as a continent). If you know your latitude and longitude, you can calculate your own grid square on the Web at **www.arrl.org/locate/grid.html**.

To earn a VUCC award, you must confirm contacts with stations in a certain number of grid squares, depending on the band.

50 or 144 MHz: 100 squares
222 or 420 MHz: 50 squares
902 or 1296 MHz: 25 squares
2300 MHz: 10 squares
3300 MHz and Up: 5 squares

You'll find out more about VUCC on the Web at **www.arrl.org/awards/vucc/**.

BREAKING THE PILEUPS

If you set your sights on one of these awards, you'll be in good company. Thousands of your fellow hams will be doing the same. When a station suddenly pops up in a rare DXCC entity, US state or grid square, the result is a chaotic swarm of signals as everyone tries to make contact. We call these on-the-air mob scenes *pileups*.

The most difficult pileup occurs when everyone, including the DX station, is on the same frequency. The hunters have no choice but to transmit over and over in hope that the DX operator will hear (or decode) their call signs.

A good DX operator will try to make order out of chaos by imposing some rules. The most common technique is to ask for calls in order of call sign district. It might sound something like this…

> **Your best chance of making contact in a pileup is when the DX station is working split.**

"J77DR QRZ for sevens only!"

This means that J77DR only wants to hear from hams in the 7th call district. If your call sign has a number other than seven, you must remain silent and wait your turn.

As pileups become massive, the only workable solution is to spread it out. DX operators do this by transmitting on one frequency while tuning and listening through a range of frequencies. See **Figure 7.1**.

For example, J77DR may transmit on 14190 kHz, but he will be listening for calls from 14195 through 14210 kHz. In this instance, the request for calls may sound like this…

"J77DR QRZ, 195 to 210!"

Or he may be somewhat less specific…

"J77DR QRZ, Up 5 to 15!"

This means that he is listening 5 to 15 kHz above his transmitting frequency.

Spreading the pileup in this fashion is known as *working split*. To make this work, you need to understand how to place your transceiver into the split frequency mode. Most modern radios have two variable frequency oscillators (VFOs) that you use to set your frequency. These separate VFOs are usually labeled A and B. When you place your radio in the **SPLIT** mode, you transmit on one VFO frequency and listen on the other. The trick is making sure that you don't have them reversed. Transmitting on the DX station's transmit frequency is a major no-no when he is working split! Instead, you want your "receive" VFO tuned to his transmit frequency while you call somewhere in his listening range with your "transmit" VFO.

Your best chance of making contact in a pileup is when the DX station is working split. You can analyze his operating patterns and pick a transmit frequency that gives you the greatest chance of being heard. Even though the

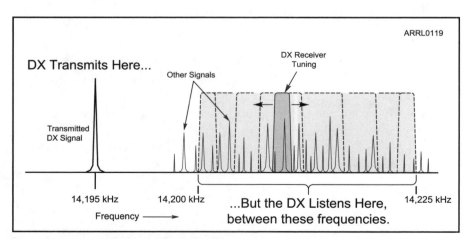

Figure 7.1—When a DX station is working split, he transmits on a single frequency, but listens through a range of frequencies. As the DX hunter, you have to work in reverse—listen on the DX station's transmit frequency and transmit somewhere in his listening range.

DXpeditions

In this chapter we mentioned that some rare DXCC entities have no hams whatsoever and are only active when other hams visit the locations. When an amateur, or group of amateurs, travel to a location with the purpose of getting on the air for a limited period of time, this is known as a *DXpedition*.

If you watch the pages of *QST* magazine, or the news stories on the ARRL Web, you'll know when DXpeditions are about to take place. Their activities are well publicized in advance because they want to contact as many people as possible.

Some DXpeditions are small—a ham may just travel to a rare location by himself and put it on the air for short period of time.

This DXpedition team traveled to rare Kure Island in the Pacific and put it on the air in 2005. Kneeling, from left: Alan, AD6E; Ward, NØAX; Ann, WA1S; Franz, DJ9ZB, and Gary, NI6T. Standing, from left: Alan, K6SRZ; John, N7CQQ; Charlie, W6KK; Arnie, N6HC; Gerd, DJ5IW; Steve, VE7CT and Bob KK6EK.

Other DXpeditions are major undertakings with many amateurs and tons of equipment. A large DXpedition to a hard-to-reach destination, such as an uninhabited, isolated island can cost hundreds of thousands of dollars! These efforts are usually financed through donations, and from the collective pockets of the DXpeditioners themselves.

It pays to keep your fingers on the pulse of DX news so that you'll know when a DXpedition is about to take to the airwaves. You may have to wait years for another chance!

DX operator says he is listening between 14190 and 14210 kHz, does he seem to favor a certain portion of that range? Listen carefully and you may detect a pattern that will give you the edge!

Here are some more pileup tips…

■ Don't call the DX unless you can hear his signal. This may sound like common sense, but you'd be surprised at how many hams will discover a pileup in progress and start throwing out their call signs when they can't even hear the DX station! This causes misery for everyone.

■ When the DX station acknowledges someone's call, stop calling until the contact is completed. The obnoxious practice of calling while the DX operator is trying to complete a contact is at the top of the list of DX annoyances. Wait until you hear "QRZ" or something similar before calling again. (In CW pileups, the end of the contact may only be signaled by the DX station sending his own call sign.)

■ When calling, send only your call sign, not the call of the DX station. And send your *complete* call sign, not just a few letters.

■ When a DX station is trying to sort out a call sign, he may send "W4? AGN," meaning that he hears a W4, but can't make out the rest of the call sign (AGN is a CW abbreviation for "again.") If you aren't the station he is looking for, don't call! This will only make him angry and might get you blacklisted.

■ If you're trying to make an SSB contact, this is probably a good time to use speech compression (or speech processing) if your transceiver offers this feature. It may give you the extra punch to get through.

■ If you make contact, send only the information the DX operator needs, typically your location and signal report. Don't try to engage in conversation; that's not what he is there for! The DX operator just wants to get you into his log and move on to the next contact.

FINDING THE DX

The most common method for finding DX is to simply sit in front of the radio and tune through the bands. Busy amateurs, however, often discover that it is difficult to set aside an hour or two to do this. If you're this busy, you may want to get some assistance from *DX Clusters*.

> **DX Clusters are networks that gather reports, called** spots, **from hams who have discovered desirable DX stations.**

DX Clusters are networks that gather reports, called *spots*, from hams who have discovered desirable DX stations. Some DX Clusters use packet radio to disseminate this information, but most hams use Internet-based clusters that they access through Web browsers or by way of telnet connections. Perhaps the most widely used Web cluster is DX Summit at **oh2aq.kolumbus.com/dxs/**.

A DX Cluster will tell you, at a glance, which DX stations have been heard on the air within the last few minutes or hours, and on which frequencies. Other information is also available such as the mode the DX station is using and whether he is working split.

Some electronic logging programs take this a step further. You can set up the program to sound an alarm if a particular DX station is spotted. Even better, the program can be configured to sound the alarm if it detects a spot for a station you need for a particular award! See the sidebar "Logging Software," earlier in the chapter.

The DX Summit DX Cluster at oh2aq.kolumbus.com/dxs/.

8 FM — "No Static at All"

Do you remember learning about FM while you studied for your license? If not, here is a refresher course in four paragraphs.

FM stands for *frequency modulation*. When you modulate a signal, you change it in a way that allows it to carry information—voices, data, images or whatever. With FM, we take a signal and modulate it by shifting its frequency back and forth.

The great advantage of FM is found in how it is received. An FM receiver *de*modulates a signal by looking for frequency shifts. Most

> **A repeater effectively listens and "talks" at the same time!**

of the noise in our environment is *not* frequency modulated. So, the FM receiver extracts the information from the FM signal and, by default, leaves out most (or all) of the noise. The result is a clean signal without static crashes, sputtering motor noises and so on. That's why FM has long been the best choice for high-fidelity audio broadcasting (although that is changing now that digital audio broadcasting and satellite radio have appeared). It is also the reason why hams enjoy using FM—signals are clear and noise is nonexistent.

For every advantage there is a disadvantage, and FM is no exception. An FM receiver requires a strong signal for *full quieting* (noise free) reception. Anything less than the required strength results in noise, and the weaker the signal, the greater the noise. In fact, it is quite difficult to understand voices when an FM signal becomes weak.

Because of this disadvantage, FM is best for local use where distances are relatively short and signals are strong. FM is a poor choice for long-distance work. In fact, it is so poor that under weak signal conditions, direct point-to-point *simplex* communication can be a challenge. With directional antennas, substantial output power and no tall obstacles between them, two stations can enjoy an FM contact with clear signals over 50 miles or more. Add buildings or hills, however, and the effective distance decreases.

The solution for limited FM range is to build an automated station with lots of power, a sensitive receiver and great antennas. Put that station on top of the nearest mountain or skyscraper (or elevate its antennas with a tall tower) and use it to relay FM signals throughout the entire area. Such Amateur Radio relay stations exist by the *thousands* throughout the United States and they are known as *repeaters*.

KC7RJK

This tower supports the antennas for the KC7RJK repeater on 145.5 MHz in Eugene, Oregon. With these highly elevated antennas, the repeater enjoys wide coverage.

FM REPEATERS

A repeater is similar to any other Amateur Radio station--it uses a transmitter, a receiver and an antenna. The magic is in the fact that the receiver and transmitter in a repeater are on different frequencies and the output of the receiver is fed to the input of the transmitter.

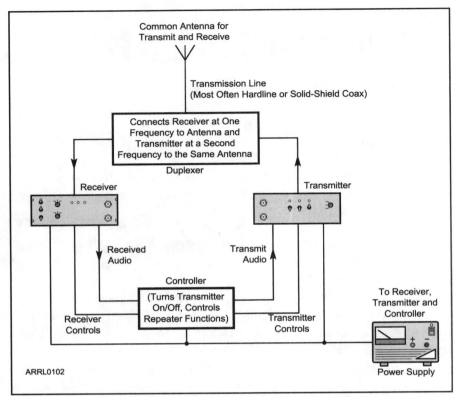

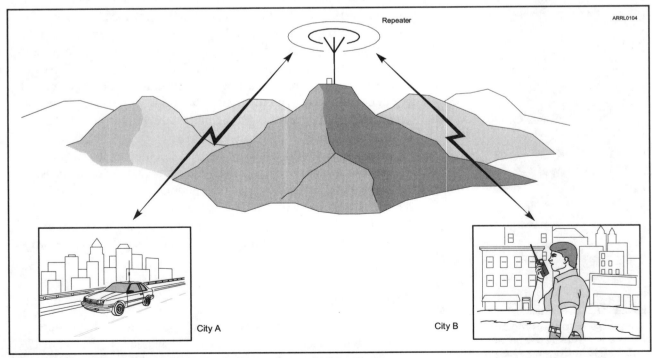

Thus, everything that the receiver hears is retransmitted *simultaneously* ("repeated") by the transmitter. A repeater effectively listens and "talks" at the same time! (See the diagram of a typical repeater in **Figure 8.1**.) In communication circles, this is known as operating *full duplex*.

Of course, your radio's receiver and transmitter are also tuned to different frequencies (the opposite of those on the repeater). Your radio transmits on the repeater's *input* frequency and receives on the repeater's *output* frequency. The same is true of the station you are talking to. The result is that the repeater is your conversational middleman—it listens to you and relays everything you say

Figure 8.1—A diagram of a typical repeater system. This repeater uses one antenna for transmitting and receiving.

Figure 8.2—A repeater on a high location, such as the mountaintop shown here, acts as a relay between two stations that wouldn't otherwise be able to hear each other.

to the other station *while you are saying it*. When it's the other station's turn to talk, the repeater does the same for him, listening to everything and retransmitting to you.

Thanks to the repeater, a conversation that might have otherwise been impossible is now possible. A repeater greatly extends the range of your radio. In the case of a handheld transceiver, it may expand it from a few miles to tens or even hundreds of miles. See **Figure 8.2**.

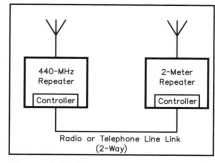

Figure 8.3—Two repeaters using different bands can be *crosslinked* as shown.

Frequency offset—the difference between the frequency on which the repeater hears and transmits—varies by repeater band. On 144 MHz the common offset is 600 kHz; on 222 MHz it's 1.6 MHz and on 450 MHz it's 5 MHz. This frequency separation is generally built in to a modern transceiver's memory so it's transparent to the user. The radio chooses the offset depending upon which band is selected.

A repeater system may also include connections to receiver and transmitter combinations on other bands. For example, a 2-meter repeater linked to the 70-cm band may receive on 147.69 MHz and transmit on 147.09, while it also receives on 449.625 and transmits on 444.625 (see **Figure 8.3**). If a signal is present on 147.69 or on 449.625, it is retransmitted on both 147.09 and 444.625. These *crosslinked* repeaters may include coverage for several bands. Such systems are capable of operating on all bands at all times, or can be set up to have remote control selectability for the various links.

There are also more complex repeater systems that form integrated wide-coverage networks. Many of these networks cover entire states and in some cases cover *several* states. They consist of linked repeaters located miles apart that are interconnected by two-way VHF or UHF links. These systems allow operators to use a local repeater to make contacts with hams in distant cities or states. As an example, the Evergreen Intertie includes more than 23 repeater stations throughout California, Oregon, Washington, Idaho, Montana, British Columbia and Alberta. An amateur on Mt Shasta can chat with a ham in Edmonton, Alberta, while both are using low-power hand-held transceivers! Integrated wide-coverage repeater systems allow users to turn links on and off as needed by using the dual-tone multifrequency (DTMF, or *TouchTone*) keypads on their radios.

LIMITING ACCESS

Most Amateur Radio repeaters are open to all users. There are no restrictions on the use of the repeater's functions. Limited-access repeaters do exist, however. Although some would argue that such operations go against the spirit of our hobby, a closed repeater is legal according to FCC regulations.

More often than not, especially in today's operating environment, you will find open repeaters that require the use of special codes or low-frequency *subaudible tones* to gain access. The reason for tone encoding the access is to prevent interference, not to limit users of the system. In cases where extraneous transmissions often activate the repeater, the use of tone encoding is the only practical way to resolve the problem. How is access to these repeaters controlled? Most often, via a

> *More often than not, especially in today's operating environment, you will find open repeaters that require the use of special codes or low-frequency subaudible tones to gain access.*

technique called *continuous tone-controlled squelch system* (CTCSS). (Many hams refer to CTCSS as *PL*—a Motorola trademark that stands for *Private Line*.) When a transmitter is configured for CTCSS, it sends a subaudible tone along with the transmitted voice or other signals. The frequency of the CTCSS tone is below the lowest audio frequency other stations will pass to their speakers, but it's sensed by a suitably equipped repeater. The repeater is programmed to respond only to carriers that send the proper tone. This effectively locks out signals that don't carry the correct CTCSS tone. Modern VHF and UHF transceivers include the necessary circuitry to generate CTCSS tones, so if you know the one you need, you can simply program it on your rig.

Alphanumeric names are used to designate the tones, and the Electronic Industries Alliance (EIA) has developed 50 standard CTCSS tone frequencies. A list of current CTCSS tones is shown in **Table 8-1**.

FINDING A REPEATER

To use a repeater, you must know one exists. There are various ways to find a repeater. Modern transceivers often include a scan mode that searches for activity. Some transceivers will even place active frequencies in their memories automatically.

There are also several very good listings (both written and software based) that can provide you with all the information available for repeaters in your area. The ARRL publishes *The ARRL Repeater Directory*, an annual, comprehensive listing of repeaters throughout the US, Canada and other parts of the world. The ARRL also publishes *TravelPlus*, a map-based CD-ROM that allows you to trace your proposed route on a color map and print a list of repeaters along the way. In addition to simply identifying local repeater activity, these directories are perfectly suited for finding repeaters during vacations and business trips. You can find more information or place an order on *ARRLWeb*: **www.arrl.org/catalog/**. Once you find a repeater to use, take some time to listen and familiarize yourself with its operating procedures.

Who Builds Repeaters?

Repeaters are expensive devices. The equipment alone can cost thousands of dollars. When you add the cost to rent space on a building or tower, and the expense of supplying electricity and possibly a telephone line or Internet connection, you're looking at serious money. Individuals can and do own repeaters, but most of these systems are sponsored and financed by ham clubs.

In most cases, you don't need to be a member of the sponsoring club to use their repeater. If you find that you are operating through a particular repeater quite a bit, however, it is a good idea to join the club that supports it. There are more than 2000 ARRL Affiliated Clubs, many of which sponsor repeaters. You can find an Affiliated Club near you by searching on the ARRLWeb at **www.arrl.org/FandES/field/club/clubsearch.phtml**.

YOUR FIRST FM TRANSMISSION

If the repeater is quiet, pick up your microphone, press the switch, and transmit your call sign.

For example, "This is W1VT monitoring."

This advises others on frequency that you have joined the system and are available to talk. After you stop transmitting, the repeater sends an unmodulated carrier for a couple of seconds to let you know it is working. Chances are that if anyone wishes to make contact they will call you at this time. Some repeaters have specific rules for making yourself heard, but usually your call sign is all you need.

It's not good repeater etiquette to call CQ. Efficient communication is the goal. You're not trying to attract the attention of someone who is casually tuning his receiver across the band. Except for scanner operation, there just isn't much tuning through the repeater bands—only listening to the machine.

If you want to join a conversation already in progress, transmit your call sign during a break between transmissions. The station that transmits next should acknowledge you. Don't use the word BREAK to join a conversation. BREAK generally suggests an emergency and indicates that all stations should stand by for the station with emergency traffic.

If you want to see if your buddy across town is on the air, call him like this:

"N1ND this is W1VT."

If the repeater is active, but the conversation in progress sounds as though it's about to end, be patient and wait until it's over before calling another station. If the conversation sounds like it's going to continue for a while, transmit your call sign between transmissions. After one of the other hams acknowledges you, politely ask to make a quick call on the repeater. Usually, the other stations will allow you this brief interruption. Make your call short. If your friend responds to your call, ask him to move to a simplex frequency or another repeater, or to stand by until the present conversation is over. Thank the other users for letting you interrupt them to place your call.

> *If the repeater is quiet, pick up your microphone, press the switch, and transmit your call sign. For example, "This is W1VT monitoring."*

Where to Operate

FM repeaters are found on many frequency bands, with the most active being on **2 meters** and **70 cm**. You'll find FM repeaters in the following frequency ranges:

51 to 54 MHz
144.60 to 145.50 MHz
146.00 to 148.00 MHz
222.25 to 224.98 MHz
442 to 445 MHz
447 to 450 MHz
918 to 921 MHz
1282 to 1288 MHz

If you're looking for direct simplex conversations, try the following:

52.525 MHz
146.52 MHz
223.52 MHz
446 MHz
1294.50 MHz

NØNQW

The Southwest Missouri SKYWARN repeater is sheltered in this unassuming structure at the base of a tall commercial tower.

Acknowledging Stations

If you're in the midst of a conversation and a station transmits its call sign between transmissions, the next station in queue to transmit should acknowledge that station and permit the newcomer to make a call or join the conversation. It's discourteous not to acknowledge him and it's impolite to acknowledge him, but not let him speak. You never know; the calling station may need to use the repeater immediately. He may have an emergency on his hands, so let him make a transmission promptly. Always remember to pause briefly at the end of your transmission (or before you jump in and respond to someone). This allows others to make themselves known.

Brevity

Always try to keep transmissions as short as possible. Short transmissions permit more people to use the repeater. All repeaters promote this practice by having timers that shut down the repeater temporarily whenever the length of a transmission exceeds a preset time limit. With this in the back of their minds, most users keep their transmissions brief. When a long-winded ham causes a repeater to shut down, this is known as *timing out* the repeater and it is usually embarrassing.

> **When a long-winded ham causes a repeater to shut down, this is known as timing out the repeater and it is usually embarrassing.**

Learn the length of the repeater's timer and stay well within its limits. The length may vary with each repeater; some are as short as 15 seconds and others are as long as three minutes. Some repeaters vary their timer length depending on the amount of traffic on frequency: the more traffic, the shorter the timer. Another purpose of a repeater timer is to prevent extraneous signals (or someone accidentally sitting on the PTT button on their mobile microphone) from holding the repeater on the air continuously. This could potentially cause damage to the repeater's transmitter.

Because of the nature of FM radio, if more than one signal is on the same frequency at one time, it creates a muffled buzz or an unnerving squawk. If two hams try to talk on a repeater at once, the resulting noise is known as a *double*. If you're in a roundtable conversation, it's easy to lose track of which station is next in line to talk. There's one simple solution to eradicate this problem forever: *Always pass off to another ham by name or call sign.* Saying, "What do you think, Jennifer?" or "Go ahead, N1TDY" will eliminate confusion and help avoid doubling. Try to hand off to whomever is next in the queue, although picking out anyone in the roundtable is better than just tossing the repeater up for grabs and inviting chaos!

The key to professional-sounding FM repeater operation is to be brisk and to the point, and to leave plenty of room for others. That's why some repeaters include a *courtesy tone* or *courtesy beep*. You'll hear it when a station stops

LEO WEHRSTEDT, VE3ATC

The repeater shown here, being tested by Laurie Bridgett, VE3BCD, is owned by the Lakehead Amateur Radio Club of Ontario, Canada. It is typical of many repeater systems. The large cylinders at the bottom are *duplexers*. They are extremely sharp filters that keep transmit and receive signals separate. Thanks to duplexers, a repeater can simultaneously receive and transmit from the same antenna without damaging sensitive circuits.

transmitting—the repeater pauses slightly, and then beeps. You are not supposed to begin talking until you hear the beep. This forces everyone to pause between transmissions to allow another station to break in. If you *do* transmit before hearing the courtesy tone, the repeater's timer won't reset. The result can be an embarrassing *time out!*

With a sensitive repeater in the area, a hand-held FM transceiver can be a blast to operate! In this example, Lori Wachtman, KC9DPC, uses an EchoLink-equipped repeater to set up a chat between her home in Fort Wayne, Indiana and a ham in Fort Carson, Colorado.

Identification

You must give your call sign at the end of each transmission or series of transmissions and at least every 10 minutes during the course of a contact. You don't have to transmit the call sign of any other station, including the one you're contacting. (Exception: You must transmit the other station's call sign when passing third-party traffic to a foreign country.)

You know it's illegal to transmit without station identification. Aside from breaking FCC rules, it's considered poor amateur practice to key your microphone to turn on a repeater without identifying your station. This is called *kerchunking* the machine. If you don't want to have a conversation, but simply want to check whether your radio works (or if you are able to access a particular repeater) just say something like "N1ND testing." This way you accomplish what you want and remain legal in doing so.

FM AND THE INTERNET

People have been enjoying voice communication over the Internet for years. It is known as Voice over Internet Protocol, or VoIP.

Hams have been putting VoIP to good use as well. Rather than relying on ionospheric propagation for long-distance communication, they've turned to VoIP in combination with VHF or UHF FM transceivers to span hundreds or thousands of miles.

There are several flavors of amateur VoIP in use today. Depending on how they are configured, these systems may involve *repeater linking* where two distant repeater systems share signals with

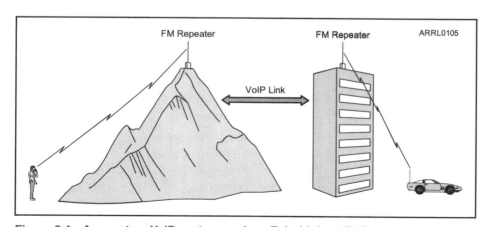

Figure 8.4—An amateur VoIP system such as EchoLink or IRLP can connect two repeaters. The repeaters could be hundreds or even thousands of miles apart.

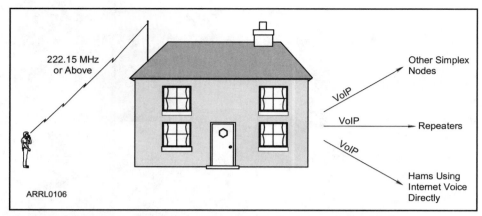

Figure 8.5—This is a diagram of a typical VoIP simplex node. In this instance, the person with the hand held transceiver is talking to his home station, which then connects him to other simplex nodes, repeaters or individual hams via the Internet.

one another (**Figure 8.4**). Another application is so-called *simplex linking* where one or more users with handheld or mobile transceivers communicate directly with a "base" station (or *node*) that is linked to the Internet (**Figure 8.5**). The one element that all amateur VoIP systems have in common is that the Internet acts as the relay between stations.

The appeal of amateur VoIP is easy to understand. Technician licensees without access to HF can use these VoIP systems to enjoy conversations with other hams far beyond the range of their FM transceivers. General and Amateur Extra hams without HF stations at home can also benefit from VoIP in the same manner.

Let's take a brief look at two of the most popular incarnations of Amateur Radio VoIP.

EchoLink

EchoLink was developed by Jonathan Taylor, K1RFD, in early 2002. In an astonishingly short period of time, EchoLink became one of the most popular Amateur Radio VoIP systems with more than 150,000 users worldwide. The free EchoLink software for *Windows* can be downloaded at **www.echolink.org**.

EchoLink can be used in several ways:

■ You can use a computer and an Internet connection to talk to hams by connecting through distant repeater systems or simplex links. In this application there is no radio at your end of the conversation.

■ You can create your own simplex link by connecting your computer to an FM transceiver at your station and using another radio (perhaps a mobile or handheld) to access the EchoLink network remotely. Many hams do this, but your range is limited by how good your base station happens to be. Also, FCC rules prohibit you from doing this on frequencies below 222.15 MHz.

■ Some repeater systems allow you to access the EchoLink network by sending a string of characters or numbers from your radio's DTMF keypad. In this way, you can "bridge" your local repeater to another EchoLink-capable repeater that is hundreds or even thousands of miles away.

"Nobody Talks to Me!"

As a new ham you may find that the repeater stays awfully quiet after you say that you are "monitoring." You've announced your presence, so why won't anyone talk to you?

Part of the problem is timing. Repeaters are most active when people are traveling to and from work in the early morning and late afternoon. At other times, and on weekends, there aren't as many stations listening.

The other issue is social. Remember that repeaters are often placed on the air by clubs and they are used for local or regional communication, usually among club members and others who have formed friendships over months and years. As the new person on the repeater, you are a stranger. Some people are reluctant to strike up conversations with people they don't know well, so they may not respond to your invitation to chat.

The solution to the timing problem is to simply use the repeater when it is most active. That's easy. The social problem is a bit harder to solve, but it *can* be solved. You need to attend club meetings. Shake some hands and introduce yourself. Ask questions (veteran hams enjoy answering questions). Volunteer to assist in club activities. By doing all these things, you'll gradually shed your "stranger" label and become part of the group, and a well-known voice on the air.

Setting up EchoLink on your station computer is easy. The software includes a "wizard" that guides you through the process. When you start the EchoLink software, your computer taps the Internet to connect to an EchoLlink server. Before you can make your first connection to the network, your call sign must be verified with the information in the FCC database. This can take minutes or hours, depending on the state of the system, but it helps reduce the chances of nonhams entering the EchoLink network.

Once you're validated (you only do this once), the rest is easy. The EchoLink server acts like a telephone switchboard in cyberspace. It maintains a directory of everyone who is connected at any moment. After browsing the directory, you can request a connection between your computer and that of another amateur.

When you connect to an individual station, the custom is to call in the same fashion as you would during a traditional on-air conversation: "W1ABC from WB8IMY." Or if you are connecting to a distant repeater system: "WB8IMY, Wallingford, Connecticut." (You need to hesitate about 2 seconds before speaking to compensate for the delay.)

The EchoLink servers also support *conferencing* where several amateurs can converse in roundtable fashion. There are even EchoLink nets that meet within these conference areas on a scheduled basis.

EchoLink Setup

If you want to enjoy EchoLink conversations while sitting at your computer, you will need a microphone headset. These are commonly available from several *QST* magazine advertisers as well as RadioShack. The microphone plug attaches to the microphone input jack of your sound card and the headphone plug typically attaches to the SPEAKER OUT jack. In addition to setting up the EchoLink software, you may also need to adjust your sound card VOLUME and RECORDING control settings in *Windows*.

If you plan to connect a radio to your computer so that you can use EchoLink over an RF link, you'll need an interface. The strong enthusiasm for EchoLink is driven by the fact that it does *not* require a specialized hardware interface for connections to transceivers. This means that you can enjoy EchoLink with the radio of your choice by using common sound card interfaces such as those sold by West Mountain Radio (the Rigblaster folks), MFJ, TigerTronics and others (do a search for these manufacturers on the Web, or see any recent issue of *QST*).

BILL GERTH, W4RK

Bill Gerth, W4RK (left) and Allen Lovett, K4XXG, operate this station at the Williamson Medical Center in Tennessee.

IRLP

With IRLP—the Internet Radio Linking Project—we enter the realm of VoIP networks that can *only* be accessed by radios.

The IRLP network consists of *nodes* on either FM repeaters or simplex. As with

EchoLink, the Internet is used as a bridge to connect IRLP nodes all over the world. By talking through an IRLP-capable repeater or simplex node, you can cruise through your neighborhood while chatting with a ham on the opposite side of the planet!

All node systems run IRLP *Linux*-based software and use specialized IRLP interfacing hardware. Individual users need only access the node and use the appropriate DTMF codes to set up an IRLP contact. Unlike EchoLink, you cannot use IRLP directly from a PC without a radio, which makes it more secure against nonham access.

To use an IRLP node you need to get the access code from the node operator or group (you may be required to join a club before you receive the access codes). Some IRLP nodes also use CTCSS subaudible tones in addition to DTMF codes to control access.

An interactive map and list of IRLP nodes is available on the Web at **status.irlp.net**. To connect to an IRLP node, you usually begin by identifying yourself and sending the DTMF access code. If you are successful, the node will respond. After that, it is a matter of stating your intentions and sending the 4-digit code for the distant node you wish to access: "WB8IMY accessing Node 5555." Once the connection is set up, you'll hear a voice ID from the target node. When you hear the confirming ID, you're free to make your call and start a conversation.

WIRES-II

WIRES-II—Wide-coverage Internet Repeater Enhancement System—is a VoIP network that is similar in function to IRLP, except that the WIRES-II node software runs under *Windows*. Like IRLP, WIRES-II is entirely radio based; you cannot access a WIRES-II node directly from the Internet. A WIRES-II host server maintains a continuously updated list of all active nodes.

There are two WIRES-II operating modes. The SRG (Sister Repeater Group) mode allows users to connect to any other WIRES-II node (up to 10 repeaters or base stations) within a group specified by the node operator. As with IRLP, DTMF tones are used to control access. Depending on how the node operator has configured his system, you may need to send a single DTMF tone before each transmission, or just at the beginning and end of your contact.

The FRG (Friends' Repeater Group) operating mode allows you to connect to any other WIRES-II node in the world. The FRG mode also allows group calling of up to 10 nodes, a kind of conferencing function. To make a regular FRG call, you press #, then five more DTMF digits depending on the ID number of the WIRES-II node you are attempting to access.

MATT ORMSBY, K7MWD

Chris, KØCAO, enjoys operating FM simplex with a portable Yagi antenna from the summit of La Plata Peak in Colorado. With this much elevation, who needs a tower?

9 Weak Signals and the World Above 50 MHz

Look beyond VHF/UHF hamming with your FM mobile or hand-held transceiver. What do you see? Is the rest of the VHF and UHF amateur spectrum a trackless desert . . . or a fertile land rich in exciting opportunities?

The answer depends on your expectations. If you expect a nonstop gabfest whenever you switch on the radio, you'll be disappointed. If you expect consistently strong signals and clear communications 24/7, you'll be disappointed again.

But . . .

If you're the type of person who enjoys taking the road less traveled, the type of person who revels in the unknown and welcomes a challenge, you've come to the right place.

"WEAK SIGNALS?"

If you listen to veteran hams talking about single sideband (SSB) and CW activity on the VHF/UHF bands, you might hear it referred to as "weak signal" operating. Doesn't sound very encouraging, does it? Where's the excitement in weak signals?

The term "weak signal" is a bit misleading. It means that you're often dealing with signals that have traveled great distances, losing much of their energy along the way. Compared to the sledgehammer signals you get from your local FM repeater, these are weak indeed. In most cases you need directional antennas to concentrate the feeble energy and perhaps a device known as a *receive preamplifier* to boost the sensitivity of your receiver. SSB and CW are used because these are among the most efficient modes when you're communicating directly on VHF and UHF. SSB and CW signals are detectable at levels where FM signals can't even be heard.

The excitement in weak-signal work is found in those moments when signals reach across great distances, farther than the so-called "line of sight" that most people think of when considering VHF or UHF. You might be listening to little more than noise until, suddenly, there is something that catches your ear. It is a whispering voice, barely understood. You move your antenna and try to get more signal. Now you can hear the voice. It is coming from a station a thousand miles away! The signal may disappear within a few minutes, so you need to jump in and try to make the contact. Your heart is pounding as you grab the microphone. Will he hear you? Can you exchange call signs and signal reports before the "magic" disappears?

The excitement in weak-signal work is found in those moments when signals reach across great distances, farther than the so-called "line of sight" that most people think of when considering VHF or UHF.

The odds of success depend on the quality of your station and your ability as an operator. Weak-signal VHF/UHF isn't easy, but then few worthwhile challenges are. If it was easy, it wouldn't be fun.

Weak-signal is also more financially demanding. It costs more to set up a good station for weak-signal work than it does for FM alone.

BUILDING A WEAK SIGNAL STATION

You will definitely need a transceiver that can operate in SSB and CW on the VHF bands. As we discussed in Chapter 1, those *multimode* radios are readily available. A dc-to-daylight transceiver (a radio that operates from 1.8 to 440 MHz) may be a worthwhile investment. You'll pay at least $1000 and probably more, but the radio will serve you well for many years and will "grow" with your new license privileges. On the other hand, if you think you'll be a dedicated VHF/UHF explorer with no future interest in HF operating, consider a dedicated VHF/UHF transceiver.

Preamplifiers

A few paragraphs ago we mentioned the receive preamplifier. This little device can be very important for weak signal work. Remember that you are dealing with signals that have traveled a considerable distance to reach you. Also recall that the feed line between your radio and your antenna will burn up some of that precious signal energy before it can arrive at your radio. With a preamplifier installed at your antenna, you can give those signals a substantial boost before they enter the feed line. Sometimes a good preamplifier can make the difference between hearing a station and hearing nothing.

> *Sometimes a good preamplifier can make the difference between hearing a station and hearing nothing.*

Preamplifiers are rated by the amount of amplification they provide (their gain, expressed in decibels or dB) and by how much extraneous noise they generate (their noise figure, or NF, which is also expressed in dB).

Yes indeed, preamps cause noise. In fact, everything causes noise in electronic circuits. It is a fact of life and the best you can do is minimize it. That's why you always want to choose the "quietest" preamplifier you can afford.

For example, a good preamplifier for the 2-meter band might have a gain of 15 dB and a noise figure of less than 1 dB. When shopping for preamplifiers, look for the highest gain and the lowest noise.

Because you'll be installing your preamplifier at the antenna, look for models that are weatherproof. You'll also need a means to get power to your preamplifier along with a method of switching it out of your feed line when you transmit (you don't want to pump 100 watts into a sensitive preamplifier!).

You can run a separate set of wires to the preamplifier to provide dc power. Another method is to send the dc power through the feed line using a dc *power inserter*. You place one inserter in your station and connect it to your dc power supply. This is where the voltage enters the feed line. Another inserter is placed outside at your preamplifier to "receive" the power. See **Figure 9.1**.

Some transceivers have power-insertion capability already built in. It usually manifests itself as a PREAMP button on the front panel, but there is a catch. In most cases this button activates a preamplifier that is built into the radio. That's a nice feature, but you really need the preamplifier to be at the antenna, not in the radio. What you want is a PREAMP button that places 13.8 V on the feed line, just like a dc power inserter. It isn't a common feature, but you may see it in dedicated VHF/UHF transceivers.

The easiest way to switch the preamplifier out of the feed line is to do it automatically. You'll find many preamplifiers that have built-in RF switching circuits. These circuits sense when you've

A weatherproof receive preamplifier installed outdoors at the antenna.

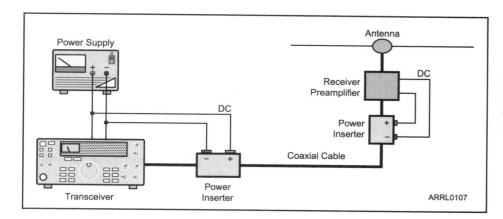

Figure 9.1—A power inserter can be used to send dc power along the coaxial cable to the antenna to power a receive preamplifier.

applied RF power and instantly switch the preamplifier out of harms way. When shopping, be sure to look for preamplifiers that offer this feature at the power level you'll be using. Don't buy an RF-switching preamplifier that is only rated for 25 watts if you plan to run 100 watts.

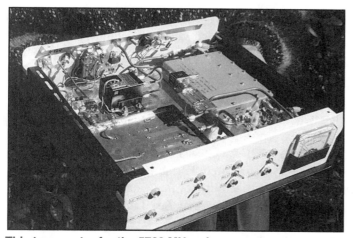

This transverter for the 5760 MHz microwave band was built by Steve Hayman, ZL1TPH. It is designed to work with any compatible 2-meter multimode transceiver.

TRANSVERTERS

An alternative to buying one or more VHF transceivers is to buy or build a *transverter* to accompany a less expensive HF-only transceiver. A transverter takes the RF output from the HF transceiver and uses it to create a signal on a particular VHF or UHF band. The transverter also converts received VHF and UHF signals to HF frequencies. In effect, a transverter turns your HF transceiver into a VHF or UHF transceiver.

Although this equipment requires some effort to interface with an HF rig (except for those made to go with your particular transceiver), the performance and cost savings can be substantial.

There are several transverter manufacturers who market to US hams, and you'll find their advertisements in *QST* magazine. Two of the most popular are Down East Microwave (**www.downeastmicrowave.com**) and SSB Electronic (**www.ssbusa.com**).

ANTENNA SYSTEMS

You'll need horizontally polarized directional antennas for best weak-signal performance. In most cases this means a multi-element Yagi or quad antenna. The more elements the antenna uses, the more tightly focused the signal will be. Of course, with a directional antenna you will also need the means to turn it, which brings the antenna rotator into play. Fortunately, most VHF/UHF directional antennas are relatively small and lightweight. You won't need a costly heavy duty rotator to turn them.

What about antenna height? Height is very important for weak-signal hamming.

Unless you already live atop a hill or mountain, you will need to elevate your antenna. That means buying and installing a tall mast or tower, or placing the antenna on your roof. There is no hard-and-fast rule for antenna height except this one: *Higher is always better*. You need to get your antenna as far above trees and other obstructions as possible. Trees are particularly bad because their leaves are filled with water and water absorbs VHF and UHF signal energy.

Aluminum towers are marvelous things. Nothing else does such a fine job of elevating your antennas. A tower is versatile; you can hang several antennas from a single tower.

But there is a downside to the tower game. Cost is a big factor. A 100-foot tower will set you back a few thousand dollars by the time you get it delivered and installed. Setting up such a beast is not a trivial exercise. You need to dig holes, pour concrete, set guy wires and more. This is an operation that requires assistance from several people; it isn't a one-person job. A building permit may be required, along with inspection and certification from a professional engineer.

If you don't see a tower in your future, look to your roof instead. There are tripod roof mounts from dealers such as RadioShack that will allow you to install an antenna rotator and a couple of antennas. Depending on the height of your roof and your overall elevation above sea level, this can be very effective—and at *much* less cost and effort. Resist the temptation to use your chimney as an antenna support. The wind mechanically loads the antenna like a sail on a boat. That loading is instantly transferred to the chimney and if the mortar is loose, you'll have bricks tumbling into your yard!

Hilltopping and Portable Operation

What if you can't put up a tower? What if your rooftop is out of the question, too? Worse still, what if you live in a valley where hillsides block your signals no matter how tall your antennas may be?

The solution is to head for the hills! VHF/UHF antennas are relatively small, and station equipment can be packed up and easily transported. Portable operation, commonly called *hilltopping* or *mountaintopping*, is a favorite activity for many amateurs. If you're on a hilltop or mountaintop, you'll have a very competitive signal. See Table 9-3 (see page 9-13) for a list of major VHF/UHF contests.

Start by setting up on an easily accessible hill or mountain for an afternoon during a contest period. For a first effort, just take along a 2-meter rig. Even if you have an FM-only rig, you can still participate. Use the common simplex frequencies, like 146.52 and 146.55 MHz. If you find that the location "plays," you know where to take your new multimode rig next time!

OUT OF THE WOODWORK

One of the secrets to enjoying SSB and CW operating on VHF and UHF is knowing how to search for contacts. Unless there is a contest going on, you're unlikely to find a conversation the moment you flip the POWER switch. There's too much space and too few operators. To make the contacts, you have to work smart.

The first step is knowing how everyone else is operating, and to follow their lead. Essentially, this means to listen first. Pay

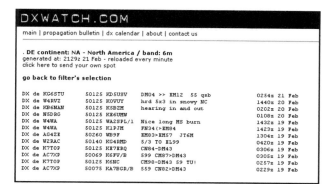

DXWATCH.COM
main | propagation bulletin | dx calendar | about | contact us

. DE continent: NA - North America / band: 6m
generated at: 2129z 21 Feb - reloaded every minute
click here to send your own spot

go back to filter's selection

DX de KG6STU   50125 KD5USV   DM04 >> EM12 55 qsb   0254z 21 Feb
DX de W4RVZ   50125 KOVUY   hrd 5x3 in snowy NC   1440z 20 Feb
DX de KB6NAN   50125 K5B2H   hearing in and out   0202z 20 Feb
DX de N5DRG   50125 KR6UMW   0108z 20 Feb
DX de W4WA   50125 WA2SPL/1 Nice long MS burn   1432z 19 Feb
DX de W4WA   50125 K1PJH   FN34(>EM84   1423z 19 Feb
DX de AC4ZE   50260 WB9F   EM83>EM57 JT6M   1304z 19 Feb
DX de W2RAC   50140 KG4RMD   5/3 TO EL99   0420z 19 Feb
DX de K7TOP   50125 KE7ERQ   CN84-DM43   0306z 19 Feb
DX de AC7XP   50069 K6FV/B   599 CM87>DM43   0305z 19 Feb
DX de K7TOP   50125 K6NC   CM98-DM43 59 TU!   0257z 19 Feb
DX de AC7XP   50075 KA7BGR/B 559 CN82>DM43   0229z 19 Feb

"DX Cluster" Web sites such as DX Watch allow you to monitor the activity on your favorite bands at your leisure.

Table 9-1

VHF/UHF Calling Frequencies

50.125 MHz
144.20 MHz
222.10 MHz
432.10 MHz

attention to the segments of the band already in use, and follow the operating practices the experienced operators are using.

How are the Bands Organized?

In most areas of the country, everyone uses *calling frequencies* to establish contact. Then the two stations move up or down the band to chat. This way, everyone can share the calling frequency without having to listen to each other's conversations. You can easily tell if the band is open by monitoring the call signs of the stations making contact on the calling frequency. A complete list of calling frequencies is shown in **Table 9-1**.

SSB/CW activity on VHF/UHF is concentrated on the two lower VHF bands, 6 and 2 meters (50 and 144 MHz). The number of active stations on these bands is about equal. Above the 2-meter band, there are considerably more active stations on the 70-cm (432 MHz) band than any other.

On 6 meters, a *DX window* has been established to reduce interference to DX stations. Yes, *DX* stations! During years of high solar activity, 6-meter openings to the other side of the world are possible! Even during the "quiet" years 6 meters will occasionally open for contacts spanning 2000 miles or more.

The window, which extends from 50.100-50.125 MHz, is intended for DX contacts only. The DX calling frequency is 50.110 MHz. US and Canadian 6-meter operators should use the domestic calling frequency of 50.125 MHz for nonDX work. When contact is established, move off the calling frequency as quickly as possible.

Activity Nights

In some areas of the country there isn't always enough activity to make it easy to find someone. Therefore, informal *activity nights* have been established. There's a lot of variation in activity nights from place to place. Check with an active VHFer near you to find out about local activity nights. See **Table 9-2**.

Contacts *can* be made on non-activity nights as well. It may just take longer to get someone's attention.

What's that Beeping?

If you tune around on 6 meters when propagation conditions are good, you'll probably hear several *beacon* stations. Beacons send their call signs and other information in slow-speed Morse code. Other information may include their output power and antenna height. Most beacons use nondirectional antennas and relatively low power. If you can hear a beacon in a certain geographic area, you can probably work stations in that area. If you hear a beacon signal from several hundred or several thousand miles away, the band is open. Move up to the calling frequency and start calling CQ to make a contact, like this...

"CQ CQ CQ from N6ATQ Norway-Six-Alfa-Tango-Quebec calling CQ and standing by."

If you're using a directional antenna, don't forget to point it toward the location of the beacon. You'll probably need to call CQ several times before you finally grab

Table 9-2

Common Activity Nights

Band (MHz)	Day	Local Time
50	Sunday	6:00 PM
144	Monday	7:00 PM
222	Tuesday	8:00 PM
432	Wednesday	9:00 PM
902	Friday	9:00 PM
1296 and up	Thursday	10:00 PM

someone's attention. When everyone realizes that the band is open, however, many more stations will show up.

Hunting for Contacts on the Web

If you don't have time to sit in front of your radio prowling for signals, the Internet offers some excellent tools. When VHFers find activity, many of them post alerts (known as *spots*) on certain Web pages. You can connect to the Web and browse these pages to see what is going on in your area.

Two highly recommended sites are: DX Summit **oh2aq.kolumbus.com/dxs/**
DX Watch **www.dxwatch.com**

These Web pages will even allow you to sort for spots for a particular band or mode of interest (6 meter SSB, for example).

You can automate the process even further by using software that constant scans for spots of interest at these Web sites and sounds an alarm to call you to your radio. Many logging applications include this feature. You'll find logging software advertised and reviewed in the pages of *QST* magazine. There are also stand-alone "DX monitors" that keep an eye on Web sites (or Internet telnet "DX Clusters") and alert you to activity of interest. Take a look at *DX Monitor* at **www.benlo.com/dxmon.html**, or simply do a Google search (**www.google.com**) using a string of keywords such as "dx spot alert software."

GRID SQUARES

When the band is open and the contacts are being made, one of the first things you'll notice is that most conversations include an exchange of *grid squares*. Grid squares are a shorthand means of describing your general location anywhere on Earth.

Grid squares are coded with a 2-letter/2-number/2-letter code (such as FN24kp). This handy designator uniquely identifies the grid square and your exact location in latitude and longitude; no two have the same identifying code.

There are several ways to find out your own grid square identifier. The first bit of information you need is the *approximate* latitude and longitude of your station. (In most cases, the latitude and longitude of your city or town will be sufficient.) Your town engineer can provide this information, or you can go to a library and check a couple of geographic atlases. A nearby airport is another good source.

If you're fortunate enough to own a *GPS* (global positioning system) receiver, you can get precise latitude and longitude information in seconds. If you don't have one of these wonders lying around, perhaps you know a friend who does.

> *Grid squares are a shorthand means of describing your general location anywhere on Earth.*

With data in hand, you're ready to determine your grid square. Just go to the ARRL Web site at **www.arrl.org/locate/ grid.html**. On this page you can plug in the numbers and find your grid square.

The ARRL also makes a large grid square map that you can hang on your wall (see **Figure 9.2**). You can use the map to keep track of the grids you've contacted by using a marker to color in the squares you've worked. You can purchase the map from the ARRL at **www.arrl.org/catalog/?item=8977**.

The Grid Square Obsession

Other than using it as a shorthand means for determining one's location, what is the big deal about exchanging and collection grid squares?

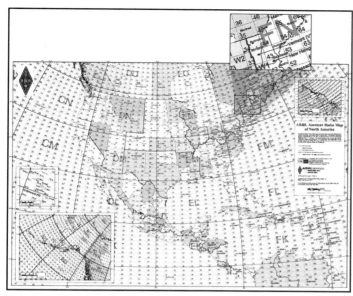

Figure 9.2—The ARRL offers this large grid-square map that you can hang on your wall. Call 888-277-5289, or order on line at www.arrl.org/catalog.

The answer, in two words, is *operating awards*.

Many hams love to work for operating awards, which are certificates or plaques that you receive for various achievements. Nearly all VHF operating awards involve the exchange of grid squares. The more unique grid squares you contact, the more awards you qualify for. See the sidebar "VHF/UHF Century Club."

The ARRL also maintains an on-line database of hams who regularly report the total number of grid squares they have confirmed on various bands. This "Standings" database is available at **www.arrl.org/qst/worldabove/ standings.html**. You can search the database to see what other hams have been doing, or add your own grid square totals!

VHF/UHF Century Club

The most popular award for VHF, UHF and microwave operating is the ARRL's VHF/UHF Century Club, or VUCC. The VUCC certificate is awarded by band. To earn a VUCC award for a given band, you need to confirm contacts with a minimum number of grid squares, depending on the band.

Band (MHz)	Grid Squares Needed for VUCC
50	100
144	100
222	50
432	50
902	25
1296	25
2300	10
Above 3400 MHz	5

For more information, go to the ARRL Web at **www.arrl.org/ awards/vucc/**.

KB8VAO received this VUCC award for contacting stations in 5 different grid squares on 24 GHz.

HOW FAR CAN YOU TALK?

If you're new to the world above 50 MHz, you might wonder what sort of range is considered "normal." To a large extent, your range on VHF is determined by your location and the quality of your station. For example, a high-power station with a stack of beam antennas on a 100-foot tower will outperform a 10-W rig and a small antenna on the roof.

But for the sake of discussion, consider a more-or-less "typical" station. On 2-meter SSB, a hypothetical typical rig would be low-powered, perhaps a multimode transceiver (SSB/CW/FM), followed by a 100-W amplifier. The antenna of our typical station might be a single 15-element Yagi at around 50 feet, fed with low-loss coax.

Using SSB or CW, how much territory could this station cover on an average evening? Location plays a big role, but it's probably safe to say you could talk to similarly equipped stations about 200 miles away almost 100% of the time. Naturally, higher-powered stations with high antennas have a greater range, up to a practical maximum of about 350-400 miles in the Midwest (less in the hilly West and East).

On 222 MHz, a similar station might expect to cover about the same distance, and somewhat less (perhaps 150 miles) on 432 MHz. This assumes normal propagation conditions and a reasonably unobstructed horizon. This range is a lot greater than you would get for noise-free communication on FM. Increase the height of the antenna to 80 feet and the range might extend to 250 miles, and probably more, depending on your location. That's not bad for reliable communication!

Band Openings and DX

The main thrill of the VHF and UHF bands is the occasional *band opening*, when signals from far away are received as if they're next door. DX of well over 1000 miles on 6 meters is commonplace during the summer, and occurs at least a few times each year on 144, 222 and 432 MHz.

DX propagation on the VHF/UHF bands is strongly influenced by the seasons. Summer and fall are definitely the most active times although band openings occur at other times as well. Here is a review of the most popular types of VHF/UHF propagation. Remember that there is a lot of variation, and that no two band openings are alike. This uncertainty is part of what makes VHF/UHF interesting and fun!

■ *Tropospheric—or simply "tropo"—openings*. Tropo is the most common form of DX-producing propagation on the bands above 144 MHz. It comes in several forms, depending on local and regional weather patterns. This is because it is caused by the weather. Tropo may cover only a few hundred miles, or it may include huge areas of the country at once. The best times of year for tropo propagation are from spring to fall, although they can occur anytime. One indicator of a possible tropo opening is dew on the grass in the evening. Another is a high-pressure weather system stalled over or near your location.

> **Tropo is the most common form of DX-producing propagation on the bands above 144 MHz.**

■ *Meteor scatter* communication uses the ionized trails meteors leave as they pass through the atmosphere. VHF radio signals can be reflected by these high-altitude meteor trails and return to Earth hundreds or even thousands of miles away (see **Figure 9.3**). This ionization lasts only a second or so. Most meteor-scatter contacts are made on 6 and 2 meters. Because the ionization from a single meteor is brief, special operating techniques are used. (See the sidebar, "Hooked on Meteors.")

Meteor-scatter contacts are possible at any time of year. Activity is greatest during the major meteor showers, especially the Perseids, which occurs in August.

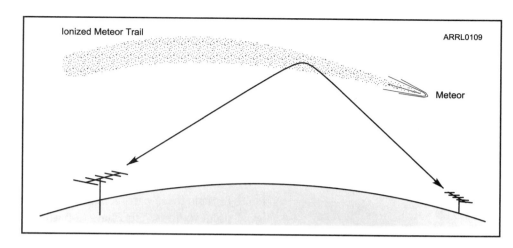

Figure 9.3—VHF radio signals can be reflected by the blazing trails of meteors as they enter our atmosphere. The results are quick contacts with stations hundreds or thousands of miles away.

■ *Sporadic E* (abbreviated E$_s$) propagation is the most spectacular DX producer on the 50-MHz band, where it may occur almost every day during late June, July and early August. A short E$_s$ season also occurs during December and January. Sporadic E is more common in mid-morning and again around sunset during the summer months, but it can occur at any time and any date. E$_s$ also occurs at least once or twice a year on 2 meters in most areas. E$_s$ results from small patches of ionization in the ionosphere's E layer. Es signals are usually strong, but they may fade away without warning. On 6 meters you can

Hooked on Meteors

It may be hard to believe, but cosmic debris is falling into our atmosphere 24 hours a day, every day. The Earth sweeps up at least 100,000 pounds of the stuff each year!

Some of the debris is sizable and falls at certain times of the year known as meteor showers. Among the largest showers is the Leonids that arrive every mid-November. As the larger particles plunge into the atmosphere, the air in from of them compacts under the extreme pressure and becomes superheated. This intense heat begins to disintegrate the meteors and we see the results as shooting stars.

As the meteors burn, they leave ionized trails that briefly act as mirrors for VHF radio signals. By bouncing your signals off these fiery trails, you can communicate up to near 2000 miles.

Hams look forward to meteor showers and attempt to make quick contacts using SSB and CW. The contacts last a few seconds at best, just long enough to trade call signs, grid squares and maybe a few more details. You can make meteor scatter contacts on the 6 meter band with relative ease. Just 100 W and an omnidirectional antenna will do the job—no directional antennas or towers required. Meteor scatter on 2 meters is more difficult, but also more popular. On 2 meters you'll need directional antennas, some height and at least 100 W output.

If you're willing to try digital communication, your meteor scatter possibilities expand tremendously. Thanks to the WSJT software suite created by Joe Taylor, K1JT, you can make meteor-scatter contacts *at any time of the day or night*. Since there are always bits of dust falling to Earth, there is no need to wait for a shower. By using digital technology, you can exploit these very brief "burns" whenever you desire.

All you need is a computer with a sound card, a VHF SSB transceiver and an interface to connect your computer to the transceiver. The interfaces are available from *QST* magazine advertisers such as MFJ, West Mountain Radio, TigerTronics, MixW Rigexpert and MicroHAM. Pick up a copy of *QST*, or search with Google for the manufacturer Web sites. The WSJT software is a free at **pulsar.princeton.edu/~joe/K1JT/**.

Most WSJT operators use a specific mode called FSK441. These contacts are not conversations per se. The idea is to send a CQ signal for 30 seconds, then listen for 30 seconds. If there is a response, you send the other station's call sign and a signal report. As soon as the other station hears your report, he also sends a report. Finally, you end with an exchange of 73s (good luck).

The WSJT software handles all the timing and decodes the received signals, using sophisticated digital signal processing to dig something intelligible out of the noise. You need to be patient with this type of communication; a digital meteor scatter exchange may take up to an hour to complete.

As with voice and CW, 6 meters is the easiest band for digital meteor scatter while 2 meters tends to be the most popular. There is a Web site dedicated to setting up contacts between digital meteor scatter operators. It is called Ping Jockey Central and you'll find it at **www.pingjockey.net**.

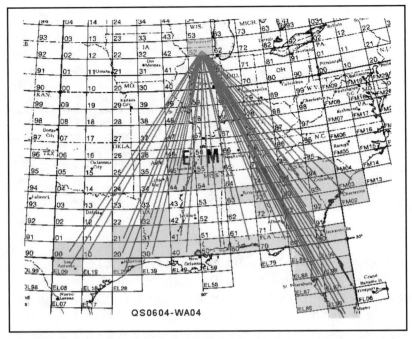

This grid square map gives you a good idea of what can happen when Sporadic E opens the band. K9NS in Illinois contacted stations in all of the gray-shaded grid squares.

make SSB Sporadic E contacts with little more than 50 W to an omnidirectional antenna!

■ *Aurora* (abbreviated Au) openings occur when the auroras are sufficiently ionized to reflect radio signals. Auroras are caused by the Earth intercepting a massive number of charged particles from the Sun. Earth's magnetic field funnels these particles into the polar regions. The charged particles often interact with the upper atmosphere enough to make the air glow. Then we can see a visual aurora. The particles also provide an irregular, moving curtain of ionization that can propagate signals for many hundreds of miles.

Aurora-reflected signals have an unmistakable ghostly sound. CW signals sound hissy; SSB signals sound like a harsh whisper. FM signals refracted by an aurora are often unreadable. (Score another one for SSB and CW!)

■ *EME, or Earth-Moon-Earth* (often called Moonbounce) is the ultimate VHF/UHF DX medium. Moonbouncers use the Moon as a reflector for their signals, and the contact distance is limited only by the diameter of the Earth (both stations must have line of sight to the Moon). See **Figure 9.4**. As you've probably guessed, Moonbouncers have a particular obsession about knowing where the Moon is, especially when they can't see it because of cloud cover.

Moonbounce conversations between the USA and Europe or Japan are commonplace—at frequencies from 50 to 10,368 MHz. That's true DX! Hundreds of EME-capable stations are now active, some with gigantic antenna arrays. Their antenna systems make it possible for stations running 100 W and one or two Yagi antennas to

Members of the Northern Lights Radio Society are proud of their portable 10 GHz microwave stations! From the left: VE3KRP, KØKFC, KMØT, WØZQ, WBØLJC, KØSHF, W9FZ, VE4MA, WØAUS, KCØP, WØLMS, NØKP, WØGHZ, KDØJI and KCØIYT.

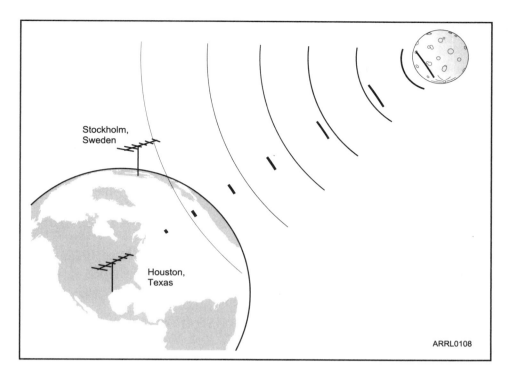

Figure 9.4—By reflecting signals off the surface of the Moon, a ham in Houston, Texas, can communicate with a ham in Stockholm, Sweden. This technique is known as moonbounce, or EME. If one station is using large antennas and high power, the other station can be very modest (100 W and a single beam antenna).

Stockholm, Sweden

Houston, Texas

ARRL0108

Digital Moonbounce

The same remarkable WSJT software that makes it possible to do meteor scatter hamming without meteor showers also makes it possible to work moonbounce without sophisticated equipment.

The JT65 and JT44 modes in the WSJT software package are specifically designed for moonbounce. Hams using little more than 100 watts to 11-element Yagi antennas have successfully made JT65/44 contacts with larger moonbounce stations. No large amplifiers needed and no towers required! Google "JT65" and "moonbounce" for more information, such as the presentation by K2UYH at **www.nitehawk.com/rasmit/jt44_50.html**.

work them. Activity is constantly increasing. In fact, the ARRL sponsors an EME contest, in which Moonbouncers compete on an international scale.

VHF CONTESTING

Amateur Radio contests test your ability to work the most stations in different geographical areas on the most bands during the contest period. Contests also give you a chance to evaluate your equipment and antennas, and to compare your results with others. In most VHF/UHF contests, each contact is worth a certain number of points. You multiply your point total by the total number of different grid squares (*multipliers*) to calculate your final score. The only restrictions in these contests are that contacts through repeaters (and satellites) don't count, and the national 2-meter FM calling frequency, 146.52 MHz, is off limits for ARRL contest contacts.

During the first hour or two of a VHF contest, contacts may come fast and furious. At other times, VHF contesting is more like an extended activity hour. VHF contests provide set times during which many other stations are operating. The concentrated activity gives you a chance for many contacts. During a contest, you'll know right away if there's a band opening!

Depending on your location, you may be able to work dozens of different grid squares on several bands, which makes for a high score and lots of fun. If you're interested in awards chasing, you'll also be pleased to know that many hams travel to rare grid squares for contests.

Table 9-3

Major VHF/UHF Contests
(See *QST* magazine for complete details.)

Contest	Bands	When?
VHF Sweepstakes	50 MHz and up	Varies according to Super Bowl date.
June VHF QSO Party	50 MHz and up	2nd full weekend
CQ Worldwide VHF	50 MHz and up	July
August UHF Contest	222 MHz and up	1st full weekend
September VHF QSO Party	50 MHz and up	2nd full weekend

Who Can Enter?

Most VHF/UHF contests are open to any licensed amateur who wants to participate. The ARRL sponsors all the major VHF/UHF contests (see **Table 9-3**), and specific rules, descriptions of the different categories, as well as entry forms, are available from ARRL Headquarters. You don't have to be an ARRL member to participate in these contests, nor are you required to submit your logs. See complete contest rules and a handy contest calendar on the ARRL Web at **www.arrl.org/contests/**.

VHF/UHF contests feature a variety of categories among which you can choose. For single operators (those operating without assistance), entry classes in the ARRL contests include all-band, single-band, low-power portable, and one for Rovers (those who operate from more than one grid square during the contest).

Microwaves

Throughout most of this chapter we've discussed weak-signal Amateur Radio as it exists on the VHF and UHF bands. As challenging as that can be, the challenge increases considerably when you begin to explore the realm of microwaves.

The microwave region begins at 1 GHz, so the lowest Amateur Radio microwave band is at 1.2 GHz. Each band has unique characteristics, but there are some rules of thumb that generally apply…

● The higher the frequency, the shorter the range.
● Directional antennas are mandatory
● Antenna height is absolutely critical. The more the better.
● Off-the-shelf equipment can be purchased for use on the 1.2, 2.3 and 10 GHz bands. Operation on other bands will probably require you to modify nonham gear.
● High power is not necessarily required. Some microwave stations transmit at less than 1 watt.

If you thought feed line loss was bad on VHF and UHF, it is *horrendous* at microwave frequencies. For this reason, it is common to use a transverter that is installed right at the antenna. The transverter will convert the received signal to a lower frequency (usually 28, 144 or 432 MHz) that will travel along the coaxial cable to your radio with minimal loss. When transmitting, process occurs in reverse—the radio sends a lower-frequency signal to the transverter and the transverter changes it to a microwave signal.

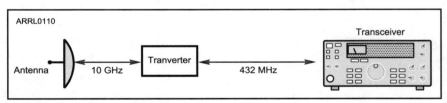

In this example, a 10 GHz transverter is being used with a VHF/UHF multimode transceiver operating at 432 MHz. The transverter converts the 432 MHz transmit signal to 10 GHz and also converts the 10 GHz received signal to 432 MHz. This keeps feed line loss to a minimum.

Microwave activity is not common. You usually won't be able to turn on your radio and make a microwave contact whenever you wish. Instead, microwave activity is usually concentrated around ARRL contests such as the June and September QSO Parties, the January VHF Sweepstakes, the UHF Contest and the 10 GHz and Up Cumulative Contest. During these events you'll often find microwave enthusiasts traveling to the nearest hills or mountains to operate portable.

Hams also prearrange contacts to test their equipment. For instance, you might ask a friend to travel to a hilltop and listen for you on 10 GHz Saturday afternoon at 3PM.

You can find more information about microwave hamming on the ARRL Web at **www.arrl.org/tis/info/microwave.html**.

Johnny, KE7V and Mike, K7MDL, set up this temporary VHF contest station atop North Point Lookout. If you want to operate from a state park, be sure to secure permission in advance and tell the rangers when you arrive.

Bruce Herrick, WW1M, captured this dramatic view of the tower and 2-meter antennas at K1WHS. Towers allow you to elevate your antennas. At VHF, UHF and microwave frequencies, higher is always better.

When and Why?

The ARRL VHF contests are held throughout the year, with emphasis on the warmer months to encourage hilltop operation. (Who wants to freeze their toes off contesting from a mountain in subzero cold?) Outside of that, the ARRL VHF/UHF contest program is designed to take the best advantage of band openings that usually occur at certain times of the year. For instance, the June VHF QSO Party almost always occurs during periods of excellent sporadic-E propagation, giving you an opportunity to enjoy long-distance contacts on 6 and 2 meters. In fact, the first documented sporadic-E contact on the 222-MHz band was made during a June VHF QSO Party.

As shown in Table 9-3, the major ARRL VHF contests consist of the January Sweepstakes, June and September VHF QSO Parties and the August UHF Contest. These events encompass many bands each. The January Sweepstakes and June and September QSO Parties are the most popular of them all, and each permit activity on SSB, CW and FM on all amateur frequencies from 50 MHz and up.

The UHF Contest is slightly different from the other contests described so far. The major difference is that only contacts on the 222-MHz and higher bands are allowed.

When to be Where

You'll find lots of random 6 and 2-meter activity during VHF contests. FM is relatively rare on 6 meters in the US, but it's quite common in most areas on 146, 222 and 440 MHz. On SSB, most stations stay near the calling frequencies of 50.125, 50.200, 144.200, 222.100 and 432.100 MHz. On CW, look between 80 and 100 kHz above 50, 144, 222 and 432 MHz. (Six meters offers less CW activity than the other VHF/UHF bands.)

> *The ARRL VHF contests are held throughout the year, with emphasis on the warmer months to encourage hilltop operation.*

Some hams who are very serious about moonbounce operating build monster antenna systems such as this one at KB8RQ in Ohio. You don't need an antenna like this to try moonbounce, however. By using the digital modes such as JT65, you can contact the big stations (like KB8RQ) with just a single Yagi antenna and 100 watts output.

THE CHALLENGE AND THE REWARD

There's no question that it's easier to get on the air with FM than SSB or CW. With FM, it may be a matter of simply buying a hand-held transceiver and talking through your local repeater. SSB and CW take a little more effort, but the reward is considerable!

As a "weak-signal" operator, you'll enjoy contacts over distances that FM enthusiasts can only achieve through complex linked-repeater systems. Best of all, you'll experience the true magic of VHF operating. As you sharpen your skills, you'll be able to predict when band openings are about to take place. By listening to the distant signals, you'll know which propagation mode is active and how to use it to your advantage.

VHF operating will challenge you every day. DX stations sometimes appear when you least expect them—and disappear just as suddenly. Wait until the day when you turn on your equipment and hear a flood of distant CW and SSB signals. The excitement will be electrifying and you'll know in that moment what you've guessed all along: there is much more to VHF than FM!

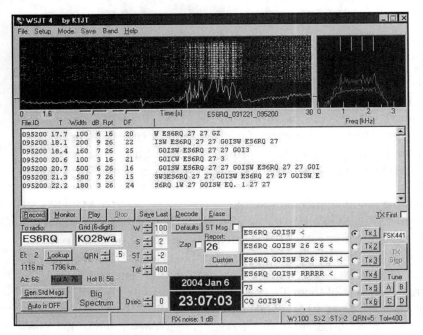

With WSJT software you can make digital meteor scatter and even moonbounce contacts with relatively inexpensive equipment. The software requires a PC running windows, a sound card and an interface to link your PC to your transceiver (you'll find these sound card interfaces in the advertising pages of *QST* magazine). The software is free and you can download it at pulsar.princeton.edu/~joe/K1JT/.

Glossary

AC hum — Unwanted 60- or 120-Hz modulation of a RF signal due to inadequate filtering in a power supply.

Adaptors — Special connectors that convert one style of connector to another.

Alternating current (ac) — Electrical current that flows first in one direction in a wire and then in the other. The applied voltage is also changing polarity. This direction reversal continues at a rate that depends on the frequency of the ac.

Amateur Satellite Corporation (AMSAT) — The organization that manages many of the amateur satellite programs.

Amateur operator — A person holding a written authorization to be the control operator of an amateur station.

Amateur Radio Emergency Service (ARES) — Sponsored by the ARRL and provides emergency communications in working with groups such as the American Red Cross and local Emergency Operations Centers.

Amateur service — A radio communication service for the purpose of self-training, intercommunication and technical investigations carried out by amateurs, that is, duly authorized persons interested in radio technique solely with a personal aim and without **pecuniary** interest.

Amateur station — A station licensed in the amateur service, including necessary equipment, used for amateur communication.

Amateur Television (ATV) — A wideband TV system that can use commercial transmission standards. ATV is only permitted on the 70-cm band (420 to 450 MHz) and higher frequencies.

Ammeter — A test instrument that measures current.

Ampere (A) — The basic unit of electrical current, also abbreviated **amps**. Current is a measure of the electron flow through a circuit. 1 Ampere is the flow of 1 **Coulomb** per second.

Amplifier — A device or piece of equipment used to increase the strength of a signal.

Amplitude modulated phone — **AM** transmission in which voice (phone) signals are used to modulate the carrier. Most AM transmission is *double-sideband* in which the signal is composed of two sidebands and a carrier. Shortwave broadcast stations use this type of AM, as do stations in the Standard Broadcast Band (535-1710 kHz). Few amateurs use double-sideband voice AM, but a variation, known as **single-sideband**, is very popular.

Amplitude modulation (AM) — The process of adding information to a signal or *carrier* by varying its amplitude characteristics.

Analog signals — A signal (usually electrical) that can have any amplitude (voltage or current) value, and whose amplitude can vary smoothly over time. Also see **digital signals**.

Antenna — A device that radiates or receives radio frequency energy.

Antenna switch — A switch used to connect one transmitter, receiver or transceiver to several different antennas.

Antenna tuner — A device that matches the antenna system input impedance to the transmitter, receiver or transceiver output impedance. Also called an *antenna-matching network, impedance matcher* or *Transmatch*.

Apogee — That point in a satellite's orbit when it is farthest from the Earth.

Automatic Position Reporting System (APRS) — A system by which amateurs can report their position automatically by radio to central servers on which their locations can be observed.

Amateur Radio Direction Finding (ARDF) — Competitions in which amateurs combine orienteering with direction finding.

Atmosphere — The mass of air surrounding the Earth. Radio signals travel through the atmosphere, and different conditions in the atmosphere affect how those signal travel or propagate.

Attenuate — To reduce the strength of a signal.

Audio frequency (AF) signal — An ac electrical signal in the range of 20 hertz to 20 kilohertz (20,000 hertz). This is called an audio signal because your ears respond to sound waves in the same frequency range.

Automatic Gain Control (AGC) — Receiver circuitry used to maintain a constant audio output level.

Automatic Level Control (ALC) — Transmitter circuitry that prevents excessive modulation of an AM or SSB signal.

Automatic control — A station operating under the control of devices or procedures that insure compliance with FCC rules.

Autopatch — A device that allows repeater users to make telephone calls through a repeater.

Balun — Contraction of "balanced to unbalanced". A device to couple a balanced load to an unbalanced feed line or device, or vice versa.

Band-pass filter (BPF) — A circuit that allows signals to go through it only if they are within a certain range of frequencies. It attenuates signals above and below this range.

Band plan — Organization of communications activity on a frequency band by general consensus.

Bandwidth — (1) Bandwidth describes the range of frequencies that a radio signal occupies. (2) FCC Part 97 defines bandwidth for regulatory purposes as "The width of a frequency band outside of which the mean power is attenuated at least 26 dB below the mean power of the transmitted signal within the band." [§97.3 (8))]

Battery — A device that converts chemical energy into electrical energy.

Battery pack — A package of several individual batteries connected together (usually in series to provide higher voltages) and treated as a single battery.

Beacon station — An amateur station transmitting communications for the purposes of observation of propagation and reception or other related experimental activities.

Beam antenna — A directional antenna. A beam antenna must be rotated to provide coverage in different directions.

Block diagram — A drawing using boxes to represent sections of a complicated device or process. The block diagram shows the connections between sections. A block diagram shows the internal functions of a complex piece of equipment without the unnecessary detail of a schematic diagram.

BNC connector — A type of connector for RF signals.

Broadcasting — Transmissions intended to be received by the general public, either direct or relayed.

Bug — A mechanical Morse key that uses a spring to send dots automatically.

Call district — The ten administrative areas established by the FCC.

Call sign — The letters and numbers that identify a specific amateur and the country in which his or her license was granted.

Calling frequency — A frequency on which amateurs establish contact and then move to a different frequency. Usually used by hams with a common interest or activity.

Capacitance — A measure of the ability of a capacitor to store energy in an *electric field*.

Capacitor — An electrical component usually formed by separating two conductive plates with an insulating material. A capacitor stores energy in an *electric field*.

CB — Citizen's Band. An unlicensed radio service operating near 27 MHz intended for use by individuals and businesses over ranges of a few miles. Also known as "11-meters" for the wavelength of its signals.

Centi — The metric prefix for 10^{-2}, or divide by 100.

Certificate of Successful Completion of Examination (CSCE) — A document that verifies that an individual has passes one or more exam elements. A CSCE is good for 365 days and may be used as evidence of having passed an element at any other amateur license exam session.

Characteristic impedance — The ratio of RF voltage and current as power moves along a feed line.

Chassis ground — The common connection for all parts of a circuit that connect to the metal enclosure of the circuit. Chassis ground is usually connected to the negative side of the power supply.

Choke filter — A type of low-pass filter that blocks RF current.

Circuit breaker — A protective component that opens a circuit or *trips* when an excessive amount of current flow occurs.

Closed repeater — A repeater that restricts access to members of a certain group of amateurs.

Closed circuit — An electrical circuit with an uninterrupted path for the current to follow. Turning a switch on, for example, closes or completes the circuit, allowing current to flow. Also called a **complete circuit**.

Coaxial cable — Coax (pronounced kó-aks). A type of feed line with one conductor inside the other and both sharing a concentric central axis.

Color code — A system in which numerical values are assigned to various colors. Colored stripes representing the different values are painted on the body of resistors and sometimes other components to show their value.

Communications emergency — A situation in which communications is required for immediate safety of human life or protection of property.

Complete circuit — An electrical circuit with an uninterrupted path for the current to follow. Turning a switch on, for example, closes or completes the circuit, allowing current to flow. Also called a **closed circuit**.

Conductor — A material whose electrons move freely in response to voltage, so an electrical current can pass through it.

Continuous wave (CW) — Radio communications transmitted by on/off keying of a continuous radio-frequency signal. Another name for international Morse code.

Control operator — An amateur operator designated by the licensee of a station to be responsible for the transmissions of an amateur station.

Control point — The locations at which a station's control operator function is performed.

Controlled environment — Any area in which an RF signal may cause radiation exposure to people who are aware of the radiated electric and magnetic fields and who can exercise some control over their exposure to these fields. The FCC generally considers amateur operators and their families to be in a controlled RF exposure environment to determine the maximum permissible exposure levels.

Coulomb (C) — The basic unit of charge. 1 Coulomb is the quantity of 6.25×10^{18} electrons. 1 Ampere equals the flow of 1 Coulomb of electrons per second.

Courtesy tone — A tone or beep transmitted by a repeater to indicate that it is okay for the next station to begin transmitting. The courtesy tone is designed to allow a pause between transmissions on a repeater, so other stations can call. It also indicates that the **time-out timer** has been reset.

CQ — "Calling any station": the general call when requesting a conversation with anyone.

Crossband — Able to receive and transmit on different amateur frequency bands. For example, a repeater might retransmit at 2 meters a signal received on 70 cm.

CTCSS — Continuous tone coded squelch system. A sub-audible tone system used on most repeaters. When added to a carrier, a CTCSS tone allows a receiver to output the received information. Also called **PL** or **sub-audible tone**.

Current — A flow of electrons in an electrical circuit.

CW (Morse code) — Radio communications transmitted by on/off keying of a continuous radio-frequency signal. Another name for international Morse code.

D region — The lowest region of the ionosphere. The D region contributes very little to short-wave radio propagation. It acts mainly to absorb energy from radio waves as they pass through it. This absorption has a significant effect on signals below about 7.5 MHz during daylight.

Data modes — Computer-to-computer communication, such as by **packet radio** or **radioteletype (RTTY)**, which can be used to transmit and receive computer files, or digital information.

DE — The Morse code abbreviation for "from" or "this is."

Deceptive signals — Transmissions that are intended to mislead or confuse those who may receive the transmissions. For example, distress calls transmitted when there is no actual emergency are false or deceptive signals.

Decibel (dB) — The smallest change in sound level that can be detected by the human ear. In electronics decibels are used to express ratios of power, voltage, or current. One dB = 10 log (power ratio) or 20 log (voltage or current ratio)

Deci — The metric prefix for 10^{-1}, or divide by 10.

Delta loop antenna — A variation of the quad antenna with triangular elements.

Detector — The stage in a receiver in which the modulation (voice or other information) is recovered from the RF signal.

Deviation — The change in frequency of an FM carrier due to a modulating signal.

Diffract — To alter the direction of a radio wave as it passes by or through the edges of obstructions such as buildings or hills.

Digital communications — see **data modes**.

Digital signal — (1) A signal (usually electrical) that can only have certain specific amplitude values, or steps—usually two; 0 and 1 or ON and OFF. (2) On the air, a digital signal is the same as a **data signal**.

Diode — An electronic component that allows electric current to flow in only one direction.

Dipole antenna — An antenna consisting of two symmetrical linear halves. Also see **Half-wave dipole**. A dipole need not be ½ wavelength long, nor is it required to have a feed point in the middle.

Direct current (dc) — Electrical current that flows in one direction only.

Direct detection — A type of **RF interference** caused by a device being disrupted by the presence of an RF signal it is not intended to receive.

Directional wattmeter (see **Wattmeter**)

Director — A parasitic element in front of the driven element in a directional antennas.

Distress call — A transmission made in order to attract attention in an emergency. (See also **MAYDAY** and **SOS**)

Doppler shift — A change in observed frequency of a signal caused by relative motion between the transmitter and receiver. Your ears hear Doppler shift when a car or train drives past you and you hear the pitch of the engine noise change. You will have to adjust your receive frequency to hear a satellite as it passes overhead because of Doppler shift.

Doubling — The undesirable act of two or more operators transmitting at the same time on the same frequency. Both operators are usually unaware of the other's presence, sometimes during the entire transmission!

Driven element — The part of an antenna that connects directly to the feed line.

Digital Signal Processing (DSP) — The process of converting an **analog signal** to **digital** form and using a microprocessor to process the signal in some way such as filtering or reducing noise.

Downlink — The frequency or frequency range on which a satellite transmits to the ground.

Dual-band antenna — An antenna designed for use on two different amateur bands.

Dummy antenna or **dummy load** — A station accessory that allows you to test or adjust transmitting equipment without sending a signal out over the air. Also called **dummy load**.

Duplex — A mode of communications (also known as *full duplex*) in which a user transmits on one frequency and receives on another frequency simultaneously. This is in contrast to half duplex, where the user transmits at one time and receives at another time.

Duplexer — A device that allows radios on two different bands to share a single antenna. Duplexers are often used to allow a dual-band radio to use a single dual-band antenna.

Duty cycle — A measure of the amount of time a transmitter is operating at full output power during a single transmission. A lower duty cycle reduces **RF radiation** exposure for the same PEP output.

DX — Distance, distant stations, foreign countries.

E region — The second lowest ionospheric region, the E region exists only during the day. Under certain conditions, it may refract radio waves enough to return them to Earth.

Earth ground — A circuit connection to a ground rod driven into the Earth or to a metallic cold-water pipe that goes into the ground.

Earth station — An amateur station located on, or within 50 km of, the Earth's surface intended for communications with space stations or with other Earth stations by means of one or more other objects in space.

Earth-Moon-Earth (EME) or **Moonbounce** — A method of communicating with other stations by reflecting radio signals off the Moon's surface.

Echolink — A system of linking repeaters and computer-based users by using the Voice-Over-Internet Protocol.

Electric field —An electric field exists in a region of space if an electrically charged object placed in the region is subjected to an electrical force.

Electromagnetic wave — A wave of energy composed of an electric and magnetic field.

Electromotive force (EMF) — The force or pressure that pushes a current through a circuit.

Electron — A tiny, negatively charged particle, normally found in the volume surrounding the nucleus of an atom. Moving electrons make up an electrical current.

Electronic keyer — A device that makes it easier to send well-timed Morse code. It sends either a continuous string of dots or dashes, depending on which lever of the *paddle* is pressed.

Element — The conducting part or parts of an antenna designed to radiate or receive radio waves.

Elmer — A ham radio mentor or teacher.

Emergency — A situation where there is a danger to human life or property.

Emergency communications — Communications conducted under adverse conditions where normal channels of communications are not available.

Emergency traffic — Messages with life and death urgency or requests for medical help and supplies that leave an area shortly after an emergency.

Emission — The transmitted signal from an amateur station.

Emission privilege — Permission to use a particular emission type (such as Morse code or voice).

Emission types — Term for the different modes authorized for use on the Amateur Radio bands. Examples are CW, SSB, RTTY and FM.

Encoding — Changing the form of a signal into one suitable for storage or transmission. *Decoding* is the process of returning the signal to its original form.

Encryption — Changing the form of a signal into a privately-known format intended to obscure the meaning of the signal. *Decryption* is the process of reversing the encoding.

Energy — The ability to do work; the ability to exert a force to move some object.

Extended-coverage receiver — A receiver that tunes frequencies from around 30 MHz to several hundred MHz or into the GHz frequencies. Also known as a **wide-range receiver**.

F region — A combination of the two highest ionospheric regions, the F1 and F2 regions. The F region refracts radio waves and returns them to Earth. Its height varies greatly depending on the time of day, season of the year and amount of sunspot activity.

False or deceptive signals — Transmissions that are intended to mislead or confuse those who may receive the transmissions. For example, distress calls transmitted when there is no actual emergency are false or deceptive signals.

Farad (F) — The basic unit of capacitance.

Federal Communications Commission (FCC) — Federal agency in the United States that regulates use and allocation of the frequency spectrum among many different services, including Amateur Radio.

Federal Registration Number (FRN) — An identification number assigned to individual by the FCC to use when performing license modification or renewal.

Feed line — The wires or cable used to connect a transmitter, receiver or transceiver to an antenna. The feed line connects to an antenna at its feed point. Also see **Transmission line**.

Feed point — The point at which a feed line is electrically connected to an antenna.

Feed point impedance — The ratio of RF voltage to current at the feed point of an antenna.

Ferrite — A ceramic material that absorbs RF energy. Ferrite is usually formed into beads or cores so that it may be placed on cables to prevent RF signal from flowing along the cable's outer conductor.

Filter — A circuit that will allow some signals to pass through it but will greatly reduce the strength of others.

Fixed resistor — An electronic component specifically designed to oppose or control current through a circuit. The **resistance** value of a fixed resistor cannot be changed or adjusted.

Form 605 — An FCC form that serves as the application for your Amateur Radio license, or for modifications to an existing license.

Forward power — The power traveling from the transmitter to the antenna along a transmission line.

Fox hunting — Exercises in which a hidden transmitter (the fox) is located in order to test direction-finding skills. Also called a *bunny hunt*.

Frequency — The number of complete cycles of an alternating current that occur per second.

Frequency band — A continuous range of frequencies in which one type of communications is authorized. An **amateur band** is a frequency band in which amateur communications take place.

Frequency coordination — Allocating repeater input and output frequencies to minimize interference between repeaters and to other users of the band.

Frequency coordinator — An individual or group that recommends repeater frequencies to reduce or eliminate interference between repeaters operating on or near the same frequency in the same geographical area.

Frequency discriminator — A type of detector used in some FM receivers.

Frequency modulated phone — The type of signals used to communicate by voice (phone) over most repeaters. FM broadcast stations and most professional communications (police, fire, taxi) use FM. VHF/UHF FM phone is the most popular amateur mode.

Frequency modulation (FM) — The process of adding information to an RF signal or *carrier* by varying its frequency characteristics.

Frequency privilege — Permission to use a particular group of frequencies.

Front-end overload — Interference to a receiver caused by a strong signal that causes the receiver's sensitive input circuitry ("front end") to be overloaded or saturated. Front-end overload results in distortion of the desired signal and the generation of unwanted spurious signals within the receiver. See also **receiver overload**.

FRS — Family Radio Service. An unlicensed radio service that uses low-power radios operating near 460 MHz and intended for short-range communications by family members.

Fuse — A thin metal strip mounted in a holder. When too much current passes through the fuse, the metal strip melts and opens the circuit.

Gain — (1) Focusing of an antenna's radiated energy in one direction. Gain in one direction means that gain in other directions is diminished. (2) The amount of amplification of a signal in a piece of equipment, such as AF Gain (volume) or RF Gain (sensitivity).

General-coverage receiver — A receiver used to listen to a wide range of frequencies, not just specific bands. Most general-coverage receivers tune from frequencies below the AM broadcast band (550 - 1700 kHz) to around 30 MHz. (See also **extended-coverage receiver.**)

GFI — Ground-fault interrupting circuit breaker that opens a circuit when an imbalance of current flow is detected between the hot and neutral wires of an ac power circuit.

Giga — The metric prefix for 10^9, or times 1,000,000,000.

GMRS — General Mobile Radio Service. A licensed radio service operating 460 MHz intended for family businesses and members to communicate within a city or region.

Go kit — A pre-packaged collection of equipment or supplies kept at hand to allow an operator to quickly report where needed in time of need.

Grace period — The time FCC allows following the expiration of an amateur license to renew that license without having to retake an examination. Those who hold an expired license may not operate an amateur station until the license is reinstated.

Grid square — A locator in the Maidenhead Locator System.

Ground connection — A connection made to the earth for electrical safety. This connection can be made inside (to a metal cold-water pipe) or outside (to a **ground rod**).

Ground rod — A copper or copper-clad steel rod that is driven into the earth. A heavy copper wire or strap connects all station equipment to the ground rod.

Ground-plane — A conducting surface of continuous metal or discrete wires that acts to create an electrical image of an antenna. **Ground-plane antennas** require a ground-plane in order to operate properly.

Ground-wave propagation — The method by which radio waves travel along the Earth's surface.

Ham-band receiver — A receiver designed to receive only frequencies in the amateur bands.

Half-wave dipole — A basic antenna used by radio amateurs. It consists of a length of wire or tubing with a feed point at the center. The entire antenna is $\frac{1}{2}$ wavelength long at the desired operating frequency.

Hand-held radio — A VHF or UHF transceiver that can be carried in the hand or pocket.

Harmful interference — Interference that seriously degrades, obstructs or repeatedly interrupts a radio communication service operating in accordance with the Radio Regulations. [§97.3 (a) (22)]

Harmonics — Signals from a transmitter or oscillator occurring on whole-number multiples (2×, 3×, 4×, etc) of the original or *fundamental* frequency.

Health and Welfare traffic – Messages about the well-being of individuals in a disaster area. Such messages must wait for **Emergency** and **Priority traffic** to clear, and results is advisories to those outside the disaster area awaiting news from family and friends.

Henry (H) — The basic unit of inductance.

Hertz (Hz) — An alternating-current frequency of one cycle per second. The basic unit of frequency.

High frequency (HF) — The term used for the frequency range between 3 MHz and 30 MHz. The Amateur HF bands are where you are most likely to make long-distance (worldwide) contacts.

High-pass filter (HPF) — A filter designed to pass signals above a specified *cutoff* frequency, while attenuating lower-frequency signals.

Impedance — The opposition to electric current in a circuit. Impedance includes both reactance and resistance, and applies to both alternating and direct currents.

Impedance match — To adjust impedances to be equal or the case in which two impedances are equal. Usually refers to the point at which a feed line is connected to an antenna or to transmitting equipment. If the impedances are different, that is a *mismatch*.

Impedance-matcher — A device that matches one impedance level to another. For example, it may match the impedance of an antenna system to the impedance of a transmitter or receiver. Amateurs also call such devices a Transmatch, antenna-matcher or **antenna tuner**.

Inductance — A measure of the ability of a coil to store energy in a *magnetic field*.

Inductor — An electrical component usually composed of a coil of wire wound on a central core. An inductor stores energy in a *magnetic field*.

Input frequency — A repeater's receiving frequency. To use a repeater, transmit on the input frequency and receive on the **output frequency**.

Insulator — A material whose electrons do not move easily, so that an electric current cannot pass through it (within voltage limits).

Integrated circuit (IC) — A compound electronic component composed of many individual components in a single package.

Intermediate frequency (IF) — The stages in a receiver that follow the input amplifier and mixer circuits. Most of the receiver's gain and selectivity are achieved at the IF stages.

International Telecommunications Union (ITU) — The organization of the United Nations responsible for coordinating international telecommunications agreements.

Internet Repeater Linking Project (IRLP) — A system of linking repeaters by using the Voice-Over-Internet Protocol.

Ionizing radiation — Electromagnetic radiation that has sufficient energy to knock electrons free from their atoms, producing positive and negative ions. X-rays, gamma rays and ultraviolet radiation are examples of ionizing radiation.

Ionosphere — A region of electrically charged (ionized) gases high in the atmosphere. The ionosphere bends radio waves as they travel through it, returning them to Earth. Also see **sky-wave propagation**.

Isotropic antenna — An antenna that radiates and receives equally in all possible directions.

K — The Morse code abbreviation for "end of transmission" or "go ahead." Any station may respond.

Keplerian elements — Mathematical values for a satellite's orbit that can be used to compute the position of a satellite at any point in time, for any position on Earth.

Key — A manually operated switch that turns a transmitter on and off to send Morse code.

Keyer or **electronic keyer** — A piece of equipment that generates Morse code automatically.

Kilo — The metric prefix for 10^3, or times 1000.

Lightning protection — Methods to prevent lightning damage to your equipment (and your house), such as unplugging equipment, disconnecting antenna feed lines and using a lightning arrestor.

Line-of-sight propagation — The term used to describe VHF and UHF propagation in a straight line directly from one station to another.

Linear amplifier - Also known as a **linear**, a piece of equipment that amplifies the output of a transmitter, often to the full legal amateur power limit of 1500 W PEP.

Local control — Operation of a station with a control operator physically present at the transmitter.

Log — The documents or log of a station that detail operation of the station. They can be used as supporting evidence, and for troubleshooting interference-related problems or complaints.

Loop antenna — An antenna with element(s) constructed as continuous lengths of wire or tubing.

Lower sideband (LSB) — (1) In an AM signal, the sideband located below the carrier frequency. (2) The common single-sideband operating mode on the 40, 80 and 160-meter amateur bands.

Low-pass filter (LPF) — A filter designed to pass signals below a specified *cutoff* frequency, while attenuating higher-frequency signals.

Malicious (willful) interference — Intentional, deliberate obstruction of radio transmissions.

Maximum useable frequency (MUF) — The highest-frequency radio signal that will reach a particular destination using **sky-wave propagation**, or *skip*. The MUF may vary for radio signals sent to different destinations.

MAYDAY — From the French *m'aidez* (help me), MAYDAY is used when calling for emergency assistance in voice modes.

Maximum Permissible Exposure (MPE) — The maximum intensity of RF radiation to which a human being may be exposed. FCC Rules establish maximum permissible exposure values for humans to RF radiation. [§1.1310 and §97.13 (c)]

Mega — The metric prefix for 10^6, or times 1,000,000.

Memory channel — Frequency and mode information stored by a radio and referenced by an alphanumeric designator.

Meteor Scatter — Communicating by reflecting signals off of the ionized trails left by meteors in the upper atmosphere.

Metric prefixes — A series of terms used in the metric system of measurement. We use metric prefixes to describe a quantity as compared to a basic unit. The metric prefixes indicate multiples of 10.

Metric system — A system of measurement developed by scientists and used in most countries of the world. This system uses a set of prefixes that are multiples of 10 to indicate quantities larger or smaller than the basic unit.

Micro — The metric prefix for 10^{-6}, or divide by 1,000,000.

Microphone — A device that converts sound waves into electrical energy (abbreviated **MIC** or **MIKE**).

Microwave — Radio waves or signals with frequencies greater than 1000 MHz (1 GHz). This is not a strict definition, just a conventional way of referring to those frequencies.

Milli — The metric prefix for 10^{-3}, or divide by 1000.

Mixer — Circuitry that combines two signals and generates signals at both their sum and difference frequencies. Mixers are used in receivers and transmitters to convert signals from one frequency to another.

Mobile station —A radio transmitter designed to be mounted in a vehicle. A push-to-talk (PTT) switch generally activates the transmitter. Any station that can be operated on the move, typically in a car, but also on a boat, a motorcycle, truck or RV.

Mode — The combination of a type of information and a method of transmission. For example, FM radiotelephony or *FM phone* consists of using FM modulation to carry voice information.

Modem — Short for *mo*dulator/*dem*odulator. A modem changes data into audio signals that can be transmitted by radio and demodulates a received signal to recover transmitted data.

Modulate or modulation — The process of adding information to an RF signal or *carrier* by varying its amplitude, frequency, or phase.

Morse code (see **CW**).

Multiband antenna — An antenna capable of operating on more than one amateur frequency band, usually using a single feed line.

Multihop propagation — Long-distance radio propagation using several skips or hops between the Earth and the ionosphere.

Multimeter — An electronic test instrument used to measure current, voltage and resistance in a circuit. Describes all meters capable of making these measurements, such as the volt-ohm-milliammeter (VOM), vacuum-tube voltmeter (VTVM) and field-effect transistor VOM (FET VOM).

Multimode radio—Transceiver capable of SSB, CW and FM operation.

Multiple Protocol Controller (MPC) — A piece of equipment that can act as a **TNC** for several **protocols**.

N or type N connector — A type of RF connector.

National Electrical Code — A set of guidelines governing electrical safety, including antennas.

Net — An formal system of operation in order to exchange or manage information

Net control station (NCS) — The station in charge of a net.

Network — A term used to describe several digital stations linked together to relay data over long distances.

National Incident Management System (NIMS) — The method by which emergency situations are managed by US public safety agencies.

Noise blanker — A circuit that mutes the receiver during noise pulses.

Noise reduction — Removing random noise from a receiver's audio output.

Nonionizing radiation — Electromagnetic radiation that does not have sufficient energy to knock electrons free from their atoms. Radio frequency (RF) radiation is nonionizing.

Notch filter — A filter that removes a very narrow range of frequencies, usually from a receiver's audio output to remove continuous tones.

Offset frequency — The difference between a repeater's transmitter and receiver frequencies. Also known as the *repeater split*.

Ohm (Ω)— The basic unit of electrical resistance.

Ohm's Law — A basic law of electronics. Ohm's Law states the relationship between voltage (E), current (I) and resistance (R). The voltage applied to a circuit is equal to the current through the circuit times the resistance of the circuit (E = IR).

Ohmmeter — A device used to measure resistance.

Omnidirectional — An antenna that radiates and receives equally in all horizontal directions.

One-way communications — Radio signals not directed to a specific amateur radio station, or for which no reply is expected. The FCC Rules provide for limited types of one-way communications on the amateur bands. [§97.111 (b)]

Open circuit — An electrical circuit that does not have a complete path, so current can't flow through the circuit.

Open repeater — A repeater that can be used by all hams who have a license that authorizes operation on the repeater frequencies.

Operator/primary station license — An amateur license actually includes two licenses in one. The operator license is that portion of an Amateur Radio license that gives permission to operate an amateur station. The **primary station license** is that portion of an Amateur Radio license that authorizes an amateur station at a specific location. The station license also lists the call sign of that station.

Oscillate — To vibrate repeatedly at a single frequency. An **oscillator** is a device or circuit that generates a signal at a single frequency.

Output frequency — A repeater's transmitting frequency. To use a repeater, transmit on the **input frequency** and receive on the output frequency.

Packet radio — A system of digital communication whereby information is broken into short bursts. The bursts ("packets") also contain addressing and error-detection information.

Paddle — Similar to a **key** or **bug**, a paddle has a pair of contacts operated by one or two levers that is used to control an electronic **keyer** that generates Morse code automatically.

Parallel circuit — An electrical circuit in which the electrons may follow more than one path in traveling between the negative supply terminal and positive terminal.

Parallel-conductor line — A type of transmission line that uses two parallel wires spaced apart from each other by insulating material. Also known as *open-wire, ladder, or window line.*

Parasitic element — Part of a directional antenna that derives energy from mutual coupling with the driven element. Parasitic elements are not connected directly to the feed line.

Part 15 — The section of the FCC's rules that deal with unlicensed devices likely to transmit or receive RF signals.

Part 97 — The section of the FCC's rules that regulate Amateur Radio.

Peak envelope power (PEP) — The average power of an RF signal at its largest amplitude peak.

Pecuniary — Payment of any type, whether money or other goods. Amateurs may not operate their stations in return for any type of payment.

Perigee — That point in the orbit of a satellite when it is closest to the Earth.

Phase — A measure of position in time within a repeating waveform, such as a sine wave. Phase is measured in degrees or radians. There are 360 degrees or 2ð???? radians in one complete cycle.

Phase modulation (PM) — The process of adding information to a signal by varying its phase characteristics. Phase modulation is very similar to **FM** and PM signals can be received by FM receivers.

Phone — Another name for voice communications.

Phone emission — The FCC name for voice or other sound transmissions.

Phonetic alphabet — Standard words used on voice modes to make it easier to understand letters of the alphabet, such as those in call signs. The call sign KA6LMN stated phonetically is *Kilo Alfa Six Lima Mike November.*

Pico — The metric prefix for 10^{-12}, or divide by 1,000,000,000,000.

PL (see **CTCSS**) — Private Line. PL is a Motorola trademark.

Polarization — The orientation of the electrical-field of a radio wave. An antenna that is parallel to the surface of the earth, such as a dipole, produces horizontally polarized waves. One that is perpendicular to the earth's surface, such as a quarter-wave vertical, produces vertically polarized waves. An antenna that has both horizontal and vertical polarization is said to be circularly polarized.

Portable device — A radio transmitting device designed to have a transmitting antenna that is generally within 20 centimeters of a human body.

Potentiometer — Another name for a **variable resistor**. The resistance value of a potentiometer can be changed over a range of values without removing it from a circuit.

Power — The rate of energy consumption or expenditure. We calculate power in an electrical circuit by multiplying the voltage applied to the circuit times the current through the circuit (P = IE).

Power amplifier — See **linear amplifier**

Power supply — A circuit that provides a direct-current output at some desired voltage from an ac input voltage.

Preamplifier — An amplifier placed ahead of a receiver's input circuitry to increase the strength of a received signal. Preamplifier circuits are often included in a receiver and may be turned on or off. Preamplifiers for VHF, UHF, and microwave frequencies are sometimes located at the antenna to amplify signals before loss in the feed line reduces their strength.

Prefix — The leading letters and numbers of a call sign that indicate the country in which the call sign was assigned.

Primary service — When a frequency band is shared among two or more different radio services, the primary service is preferred. Stations in the **secondary service** must not cause harmful interference to, and must accept interference from stations in the primary service. [§97.303]

Primary station license — An amateur license actually includes two licenses in one. The **operator license** is that portion of an Amateur Radio license that gives permission to operate an amateur station. The primary station license is that portion of an Amateur Radio license that authorizes an amateur station at a specific location. The station license also lists the call sign of that station.

Priority traffic — Emergency-related messages, but not as important as **Emergency traffic**.

Procedural signals (prosign) — For Morse code communications, one or two letters sent as a single character. Amateurs use prosigns in CW contacts as a short way to indicate the operator's intention. Some examples are K for "Go Ahead or AR for "End of Message." (A bar over the letters indicates that the prosign is sent as one character.) For phone communications, words such as "Break" or "Over" that control the flow of the communications.

Product detector — A type of mixer circuit that allows a receiver to demodulate CW and SSB signals.

Propagation — How radio waves travel.

Protocol — A method of encoding, packaging, exchanging, and decoding digital data.

Push to talk (PTT) — Turning a transmitter on and off manually with a switch, usually thumb- or foot-activated.

Q signals — Three-letter symbols beginning with Q used on CW to save time and to improve communication. Some examples are QRS (send slower), QTH (location), QSO (ham conversation) and QSL (acknowledgment of receipt).

Q system — A method of providing signal quality reports on a scale of 1 ("Q1") to 5 ("Q5").

QSL card — A postcard that serves as a confirmation of communication between two hams. QSL is a Q-signal meaning "received and understood."

QSO — A conversation between two radio amateurs. QSO is a Q-signal meaning "I am in contact."

Quad antenna — An antenna built with its elements in the shape of four-sided loops.

Quarter-wavelength vertical antenna — An antenna constructed of a quarter-wavelength long radiating element placed perpendicular to the earth.

Radiation pattern — A graph showing how an antenna radiates and receives in different directions. An *azimuthal pattern* shows radiation in horizontal directions. An *elevation pattern* shows how an antenna radiates and receives at different vertical angles.

Radio Amateur Civil Emergency Service (RACES) —A part of the Amateur Service that provides radio communications for civil preparedness organizations during local, regional or national civil emergencies.

Radio frequency (RF) exposure — FCC Rules establish maximum permissible exposure (MPE) values for humans to RF radiation. [§1.1310 and §97.13 (c)]

Radio frequency (RF) radiation or **waves** — Electromagnetic energy that travels through space without wires.

Radio frequency (RF) signals — RF signals are generally considered to be any electrical signals with a frequency higher than 20,000 Hz, up to 300 GHz.

Radio-frequency interference (RFI) — Disturbance to electronic equipment caused by radio-frequency signals.

Radiogram — A formal message exchanged via radio.

Radio horizon — The most distant point to which radio signals can be sent directly without reflections.

Radioteletype (RTTY) — Radio signals sent from one teleprinter machine to another machine. Anything that one operator types on his teleprinter will be printed on the other machine. Also known as narrow-band direct-printing telegraphy.

Ragchew — An informal conversation

Range — The longest distance over which radio signals can be exchanged.

Receiver—A device that converts radio waves into signals we can hear or see (abbreviated **RCVR**).

Receiver overload — Interference to a receiver caused by a RF signal too strong for the receiver input circuits. A signal that overloads the receiver RF amplifier (front end) causes front-end overload. Receiver overload is sometimes called RF overload.

Reciprocal operating authority — Permission for amateur radio operators from another country to operate in the US using their home license. This permission is based on various treaties between the US government and the governments of other countries.

Rectifier — A diode intended for use with high current or voltage in power supplies.

Reflected power — The power that returns to the transmitter from the antenna along a transmission line.

Reflection — Signals that travel by **line-of-sight propagation** are reflected by large objects like buildings.

Reflector — A parasitic element behind the driven element in a directional antennas.

Refract — Bending of an electromagnetic wave as it travels through materials with different properties. Light refracts as it travels from air into water. Radio waves refract as they travel through the ionosphere. If the radio waves refract enough they will return to Earth. This is the basis for long-distance communication on the **HF bands**.

Region — One of the three administrative areas defined by the **ITU**.

Remote control — Operation of a station in which the control functions of the station are operated by a control operator over a control link.

Repeater station — An amateur station that receives the signals of other stations and retransmits them for greater range.

Resistance — The ability to oppose an electric current.

Resistor —An electronic component specifically designed to oppose or control current through a circuit.

Resonance — The condition of an applied signal or wave having the same frequency as the resonant frequency of a tuned circuit or antenna.

Resonant frequency — The desired operating frequency of a tuned circuit. In an antenna, the resonant frequency is one where the feed-point impedance is composed only of resistance.

RF burn — A burn produced by coming in contact with exposed RF voltages.

RF carrier — A steady radio frequency signal that is modulated to add an information signal to be transmitted. For example, a voice signal is added to the RF carrier to produce a **phone emission** signal.

RF feedback — Distortion caused by RF signals disturbing the function of an audio circuit.

RF overload — Another term for receiver overload.

RF safety — Preventing injury or illness to humans from the effects of radio-frequency energy.

Rig—The radio amateur's term for a transmitter, receiver or transceiver.

RST — A system of numbers used for signal reports: R is readability, S is strength and T is tone. (On single-sideband phone, only R and S reports are used.)

Rubber duck antenna — A flexible rubber-coated antenna that is inexpensive, small, lightweight and difficult to break. Rubber ducks are used mainly with hand-held VHF or UHF transceivers.

S meter — A meter that provides an indication of the relative strength of received signals.

Safety interlock — A switch that automatically turns off power to a piece of equipment when the enclosure is opened.

Scattering — Radio wave propagation by means of multiple reflections in the layers of the atmosphere or from an obstruction.

Schematic diagram — A drawing that describes the electrical connections in a piece of electric or electronic equipment.

Schematic symbol — A standardized symbol used to represent an electrical or electronic circuit component on a schematic diagram.

Secondary service — When a frequency band is shared among two or more different radio services, the **primary service** is preferred. Stations in the secondary service must not cause harmful interference to, and must accept interference from stations in the primary service. [§97.303]

Selectivity — The ability of a receiver to distinguish between signals. Selectivity is important when many signals are present and when it is desired to receive weak signals in the presence of strong signals

Sensitivity — The ability of a receiver to detect weak signals.

Series circuit — An electrical circuit in which all the electrons must flow through every part of the circuit because there is only one path for the electrons to follow.

Shack — The room where an Amateur Radio operator keeps his or her station equipment.

Shielding — Surrounding an electronic circuit to block RF signals from being radiated or received.

Short circuit — An electrical circuit in which the current does not take the desired path, but finds a shortcut instead. Often the current flows directly between the negative power-supply terminal and the positive one, bypassing the rest of the circuit.

Sidebands — The sum or difference frequencies generated when an RF carrier is mixed with an audio signal. Single-sideband phone (SSB) signals have an upper sideband (USB — that part of the signal above the carrier) and a lower sideband (LSB — the part of the signal below the carrier). SSB transceivers allow operation on either USB or LSB.

Signal generator — A device that produces a low-level signal that can be set to a desired frequency.

Signal report — An evaluation of the transmitting station's signal and reception quality.

Simplex operation — Receiving and transmitting on the same frequency.

Single sideband (SSB) phone —SSB is a form of double-sideband amplitude modulation in which one sideband and the carrier are removed. SSB is a common mode of voice operation on the amateur bands.

Skip — Propagation by means of ionospheric reflection. Traversing the distance to the ionosphere and back to the ground is called a *hop*.

Skip zone — An area of poor radio communication, too distant for ground waves and too close for sky waves.

Sky-wave propagation — The method by which radio waves travel through the ionosphere and back to Earth. Sometimes called *skip*, sky-wave propagation has a far greater range than **line-of-sight** and **ground-wave propagation**.

Slow-Scan Television (SSTV) — A television system used by amateurs to transmit pictures within a signal bandwidth allowed on the HF or VHF/UHF bands by the FCC. It takes approximately 8 seconds to send a single black and white SSTV frame, and between 12 seconds and 4½ minutes for the various color systems currently in use on the HF bands.

SMA connector — A type of RF connector

SOS — A Morse code call for emergency assistance.

Space station — An amateur station located more than 50 km above the Earth's surface.

Specific absorption rate (SAR) — A term that describes the rate at which RF energy is absorbed into the human body. Maximum permissible exposure (MPE) limits are based on whole-body SAR values.

Speech compression or **processing** — Increasing the average power of a voice signal by amplifying low-level components of the signal more than high-level components.

Splatter — A type of interference to stations on nearby frequencies. Splatter occurs when a transmitter is overmodulated.

Sporadic E — A form of enhanced radio-wave propagation that occurs when radio signals are reflected from small, dense ionization patches in the E region of the ionosphere. Sporadic E is observed on the 15, 10, 6 and 2-meter bands, and occasionally on the 1.25-meter band.

Spurious emissions — Signals from a transmitter on frequencies other than the operating frequency.

Squelch — Circuitry that mutes an FM receiver when no signal is received.

SSB — Abbreviation for the **single sideband phone** mode of communication. This is the most widely used mode for phone operation on the HF bands.

Standard frequency offset — The standard transmitter/receiver frequency offset used by a repeater on a particular amateur band. For example, the standard offset on 2 meters is 600 kHz. Also see **Offset frequency**.

Standing-wave ratio (SWR) — Sometimes called voltage standing-wave ratio (VSWR). A measure of the impedance match between the feed line's characteristic impedance and the load (usually an antenna). Also, with a Transmatch in use, a measure of the match between the feed line from the transmitter and the antenna system. The system includes the Transmatch and the line to the antenna. VSWR is the ratio of maximum voltage to minimum voltage along the feed line, also the ratio of antenna impedance to feed-line impedance.

Station grounding — The practice of connecting all station equipment to a good earth ground to improve both safety and station performance.

Station license — An amateur license actually includes two licenses in one. The **operator license** is that portion of an Amateur Radio license that gives permission to operate an amateur station. The primary station license is that portion of an Amateur Radio license that authorizes an amateur station at a specific location. The station license also lists the call sign of that station.

Station records/station log — The documents or log of a station that detail operation of the station. The log can be used as supporting evidence, and for troubleshooting interference-related problems or complaints.

Stratosphere — The part of the Earth's atmosphere that extends from about 7 miles to 30 miles above the earth. Clouds rarely form in the stratosphere.

Sub-audible tone — see CTCSS.

Suffix — The letters that follow a call sign prefix identifying a specific amateur.

Sunspot cycle — The number of **sunspots** increases and decreases in a predictable cycle that lasts about 11 years.

Sunspots — Dark spots on the surface of the sun. When there are few sunspots, long-distance radio propagation is poor on the higher-frequency bands. When there are many sunspots, long-distance HF propagation improves.

Surge protector — A device that limits voltage by changing from an insulator to a conductor when excessive voltage occurs. Surge protectors are used to prevent temporary or *transient* excessive voltages from damaging sensitive electronic equipment.

Switch — A device used to connect or disconnect electrical contacts.

SWR meter — A measuring instrument that can indicate when an antenna system is working well. A device used to measure SWR.

Tactical call signs — Names used to identify a location or function during local emergency communications.

Tactical communications — A first-response communications under emergency conditions that involves a few people in a small area.

Telecommand operation — A one-way radio transmission to start, change or end functions of a device at a distance.

Telegraph key — A telegraph key (also called a *straight key*) is the simplest type of Morse code sending device.

Teleprinter — A machine that can convert keystrokes (typing) into electrical impulses. The teleprinter can also convert the proper electrical impulses back into text. Computers have largely replaced teleprinters for amateur radioteletype work.

Television interference (TVI) — Interruption of television reception caused by another signal.

Temperature inversion — A condition in the atmosphere in which a region of cool air is trapped beneath warmer air.

Temporary state of communications emergency — When a disaster disrupts normal communications in a particular area, the FCC can declare this type of emergency. Certain rules may apply for the duration of the emergency.

Terminal — An inexpensive piece of equipment that can be used in place of a computer in a packet radio station.

Third-party — An unlicensed person on whose behalf communications is passed by amateur radio.

Third-party communications — Messages passed from one amateur to another on behalf of a third person.

Third-party communications agreement — An official understanding between the United States and another country that allows amateurs in both countries to participate in third-party communications.

Third-party participation — An unlicensed person participating in amateur communications. A control operator must ensure compliance with FCC rules.

Ticket — A common name for an Amateur Radio license.

Time-out timer — A device that limits the amount of time any one person can talk through a repeater.

Terminal Node Controller (TNC) — A device that acts as an interface between a computer and a radio. It includes a **modem** and implements the rules of a **protocol**.

T-R switch — Transmit-Receive switch. A circuit or device that switches an antenna between transmitter and receiver circuits or equipment.

Traffic — Formal messages exchanged via radio. *Traffic handling* is the process of exchanging traffic. A *traffic net* is a net specially created and managed to handle traffic.

Transceiver — A radio transmitter and receiver combined in one unit (abbreviated **XCVR**).

Transistor — A solid-state device made of three layers of semiconductor material. A transistor can be used as a switch or amplifier.

Transmission line — The wires or cable used to connect a transmitter or receiver to an antenna. Also called **feed line**.

Transmitter — A device that produces radio-frequency signals (abbreviated **XMTR**).

Trip — Activate based on some physical action (current, signal level, voltage) exceeding a threshold. A circuit breaker trips, opening a circuit, when excessive current flow occurs, for example.

Troposphere — The region in Earth's atmosphere just above the Earth's surface and below the ionosphere.

Tropospheric bending — When radio waves are bent in the troposphere, they return to Earth farther away than the visible horizon.

Tropospheric ducting — A type of VHF propagation that can occur when warm air overruns cold air (a temperature inversion).

UHF connector — A type of RF connector.

Ultra high frequency (UHF) — The term used for the frequency range between 300 MHz and 3000 MHz (3 GHz). Technician licensees have full privileges on all Amateur UHF bands.

Ultraviolet (UV) — Electromagnetic waves with frequencies higher than visible light. Literally, "above violet," which is the high-frequency end of the visible range.

Unbalanced line — Feed line with one conductor at ground potential, such as coaxial cable.

Uncontrolled environment — Any area in which an RF signal may cause radiation exposure to people who may not be aware of the radiated electric and magnetic fields. The FCC generally considers members of the general public and an amateur's neighbors to be in an uncontrolled **RF radiation** exposure environment to determine the maximum permissible exposure levels.

Unidentified communications or signals — Signals or radio communications in which the transmitting station's call sign is not transmitted.

Universal Licensing System (ULS) — FCC database for all FCC radio services.

Uplink — The frequency or frequency range on which signals are transmitted from the ground to a satellite.

Upper sideband (USB) — (1) In an AM signal, the sideband located above the carrier frequency. (2) The common single-sideband operating mode on the 20, 17, 15, 12 and 10-meter HF amateur bands, and all the VHF and UHF bands.

Vanity call — A call sign selected by an amateur instead of sequentially assigned by the FCC.

Variable resistor — A resistor whose value can be adjusted over a certain range, without removing it from a circuit.

Variable-frequency oscillator (VFO) — An oscillator used in receivers and transmitters. The frequency is set by a tuned circuit using capacitors and inductors and can be changed by adjusting the components of the tuned circuit.

Vertical antenna — A common amateur antenna whose radiating element is vertical. There are usually four or more radial elements parallel to or on the ground.

Very high frequency (VHF) — The term used for the frequency range between 30 MHz and 300 MHz. Technician licensees have full privileges on all Amateur VHF bands.

Visible horizon — The most distant point one can see by line of sight.

Voice — Any of the several methods used by amateurs to transmit speech.

Voice communications — Hams can use several voice modes, including FM and SSB.

Volt (V) — The basic unit of electrical potential or EMF.

Voltage — The EMF or electrical potential difference that causes electrons to move through an electrical circuit.

Voltmeter — A test instrument used to measure voltage.

Volunteer Examiner (VE) — A licensed amateur who is accredited by a Volunteer Examiner Coordinator (VEC) to administer amateur license examinations.

Volunteer Examiner Coordinator (VEC) — An organization that has entered into an agreement with the FCC to coordinate amateur license examinations.

Voice-Operated Transmission (VOX) — Turning a transmitter on and off under control of the operator's voice.

Watt (W) — The unit of power in the metric system. The watt describes how fast a circuit uses electrical energy.

Wattmeter — Also called a *power meter*, a test instrument used to measure the power output (in watts) of a transmitter. A directional wattmeter measures both forward and reflected power in a feed line.

Wavelength — Often abbreviated λ. The distance a radio wave travels in one RF cycle. The wavelength relates to frequency. Higher frequencies have shorter wavelengths.

Weak-signal modes — Usually SSB or CW modes, used in relation to operating on the VHF and UHF bands, where many amateurs only operate FM phone.

Whip antenna — An antenna with an element made of a single, flexible rod or tube.

Winlink — A system of email transmission and distribution using Amateur Radio for the connection between individual amateurs and mailbox stations known as *Participating Mailbox Operators (PMBO)*.

Wiring diagram — A pictorial or descriptive drawing that shows how the wiring of a piece of electric or electronic equipment is to be done.

WWV/WWVH — Radio stations run by the US NIST (National Institute of Standards and Technology) to provide accurate time and frequencies.

XCVR — Transceiver

XMTR — Transmitter

Yagi antenna — The most popular type of directional (beam) antenna. It has one driven element and one or more additional parasitic elements.

73 — Ham lingo for "best regards." Used on both phone and CW toward the end of a contact.

88 — Ham lingo for "love and kisses" (not meant literally) when concluding a contact with a female operator.

INDEX

NOTES

NOTES

NOTES

NOTES

NOTES

NOTES

NOTES

Advertisers

Alinco

Cable X-Perts, Inc.

Command Productions

Cubex Company

Dave's Hobby Shop

Ham Radio Outlet

HamTestOnline

ICOM America

Mayberry Sales & Service, Inc.

MY-LOR, Inc.

NCG Company

Powerwerx

Radioware/Radio Bookstore

TENNADYNE

Teri Software

Universal Radio Inc.

VIS Study Guides

W3FF Antennas

WBØW Inc.

Yaesu USA

D-STAR REPEATER SYSTEM

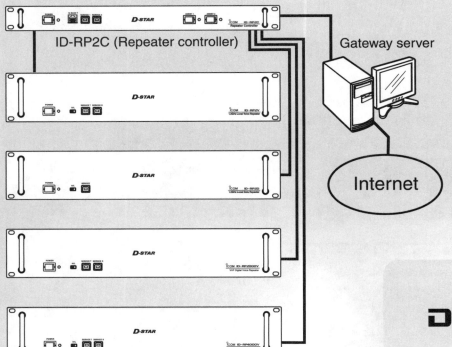

ID-RP2C (Repeater controller)

Gateway server

Internet

CONNECT TO THE INTERNET
A D-STAR Digital Voice network creates a propagation-independent worldwide amateur radio network with D-STAR digital voice repeaters via the Internet.

DIGITAL

D-STAR: JOIN THE DIGITAL REVOLUTION!

D-STAR connections are completely transparent to your laptop or other network device. Why run cables for a temporary or portable installation when mobile rigs will do the job? Connect across miles instead of meters! Icom offers a variety of digital and digital upgradeable radios. Available now!

ID-1
GO DIGITAL ON 1.2GHz!

10 Watt **I** High Speed Digital Data, Digital Voice, Analog Voice (FM) **I** Wireless Internet/Network Capable **I** PC Control via USB port **I** Digital Callsign & Digital Code Squelch

ID-800
GO DIGITAL ON 2m & 70cm!

55 Watt VHF/50 Watt UHF **I** Wide RX: 118-173, 230-549, 810-999 MHz (Cellular Blocked) **I** Analog/Digital Voice & Data **I** Callsign Squelch **I** CTCSS & DTCS Encode/Decode w/Tone Scan

IC-2200H
DIGITAL UPGRADEABLE FOR 2m!

65 Watt **I** 207 Alphanumeric Memories **I** Digital Voice & Data w/Optional UT-118 **I** Optional Callsign Squelch **I** CTCSS & DTCS Encode/ Decode w/Tone Scan **I** Weather Alert

2m and 70cm digital voice meet Main Street, U.S.A.!

ID-RP2000V
ID-RP4000V

Meet the future of amateur radio, today. D-STAR repeater systems are coming to more locations around the country. Jump in on the exciting new digital direction ham radio's taking!

D-PRS®

The low-speed data port found in Icom D-STAR handheld radios is compatible with a GPS NMEA data interface. With GPS data integrated into the D-STAR digital data stream, your location data is forwarded to the D-PRS server where gateway software connects you to the APRS® reporting system. The interface is built-in to D-STAR radios – no separate TNC and transceiver required!

DIGITAL SIGNALS = PIN-DROP CLARITY

Since D-STAR is fully digital at the terminal-to-terminal level, the network provides noise-free stable communications as if all of the D-STAR terminals are located in the same area, regardless of real geographical separation!

INTEROPERABILITY, CROSS BAND OPERATION

D-STAR digital voice radios can communicate even when they are on different bands. For example, a 2m DV radio can talk to a 70cm or 1.2GHz DV radio via repeater(s). Make D-STAR your new emergency communications standard.

THE FUTURE IS GLOBAL

Take your 2m/70cm nationwide... even worldwide! Icom D-STAR capable handhelds and mobiles, like the IC-U82 and IC-2200H shown below, now allow your local or regional communicatons to go global when used with these new D-STAR repeaters. It's a whole new ball game!

IC-V82 & IC-U82
DIGITAL UPGRADEABLE FOR 2m OR 70cm!

7 Watt VHF or 5 Watt UHF ▌200 Alphanumeric Memories ▌ Digital Voice & Data w/Optional UT-118 ▌ Optional Callsign Squelch ▌ CTCSS & DTCS Encode/Decode w/Tone Scan ▌ Weather Alert